STRATEGIES FOR E-SERVICE, E-GOVERNANCE, AND CYBERSECURITY

Challenges and Solutions for Efficiency and Sustainability

STRATEGIES FOR E-SERVICE, E-GOVERNANCE, AND CYBERSECURITY

Challenges and Solutions for Efficiency and Sustainability

Edited by

Bhaswati Sahoo, PhD (continuing)
Rabindra Narayana Behera, PhD
Sasmita Rani Samanta, PhD
Prasant Kumar Pattnaik, PhD

First edition published 2022

Apple Academic Press Inc.
1265 Goldenrod Circle, NE,
Palm Bay, FL 32905 USA

4164 Lakeshore Road, Burlington,
ON, L7L 1A4 Canada

CRC Press
6000 Broken Sound Parkway NW,
Suite 300, Boca Raton, FL 33487-2742 USA

2 Park Square, Milton Park,
Abingdon, Oxon, OX14 4RN UK

© 2022 Apple Academic Press, Inc.

Apple Academic Press exclusively co-publishes with CRC Press, an imprint of Taylor & Francis Group, LLC

Library and Archives Canada Cataloguing in Publication

Title: Strategies for e-service, e-governance, and cyber security : challenges and solutions for efficiency and sustainability / edited by Bhaswati Sahoo, PhD (continuing), Rabindra Narayana Behera, PhD, Sasmita Rani Samanta, PhD, Prasant Kumar Pattnaik, PhD.

Names: Sahoo, Bhaswati, editor. | Behera, Rabindra Narayana, editor. | Samanta, Sasmita Rani, editor. | Pattnaik, Prasant Kumar, 1969- editor.

Description: First edition. | Includes bibliographical references and index.

Identifiers: Canadiana (print) 20210287888 | Canadiana (ebook) 20210288051 | ISBN 9781771889704 (hardcover) | ISBN 9781774638811 (softcover) | ISBN 9781003131175 (ebook)

Subjects: LCSH: Computer security.

Classification: LCC QA76.9.A25 S77 2022 | DDC 005.8—dc23

Library of Congress Cataloging-in-Publication Data

....

CIP data on file with US Library of Congress

....

ISBN: 978-1-77188-970-4 (hbk)
ISBN: 978-1-77463-881-1 (pbk)
ISBN: 978-1-00313-117-5 (ebk)

About the Editors

Bhaswati Sahoo, PhD

Assistant Professor, School of Computer Engineering,
KIIT Deemed to be University, India

Bhaswati Sahoo, PhD (continuing), is Assistant Professor at the School of Computer Engineering, KIIT University, Bhubaneswar, India. Her fields of interest include e-governance, cybersecurity, and data analytics. Dr. Sahoo has published a number of research papers in peer-reviewed international journals and conferences. She has completed an MTech in computer science and engineering and is currently pursuing a PhD.

Rabindra Narayana Behera, PhD

Senior Technical Director, National Informatics Center,
Odisha State Centre, Bhubaneswar, India

Rabindra Narayana Behera, PhD, is the Senior Technical Director (Scientist-F) at the National Informatics Centre at Odisha State Centre, Bhubaneswar, India. He has more than 25 years of experience in the area of IT. He has coordinated various governance portals at state and national levels. His research area includes cloud computing, data analytics, e-governance, cybersecurity, and web services. He has published many research papers in peer-reviewed international journals and conferences. Dr. Behera's PhD is on Computer Science in the field of Artificial Intelligence.

Sasmita Rani Samanta, PhD

Professor and Pro-Vice Chancellor, KIIT Deemed To be University,
Bhubaneswar, India

Sasmita Rani Samanta, PhD, is Professor in Commerce and Management at KIIT Deemed to be University, Bhubaneswar, India, where she also holds the post of Pro-Vice Chancellor. She has more than a decade of teaching and research experience. Dr. Samanta has published many research articles in the field of e-governance, e-marketing, and brand management in reputed international journals and conferences. Dr. Samanta's PhD degree is in management science.

Prasant Kumar Pattnaik, PhD

Professor, School of Computer Engineering,
KIIT Deemed to be University, Bhubaneswar, India

Prasant Kumar Pattnaik, PhD, is Professor at the School of Computer Engineering, KIIT Deemed to be University, Bhubaneswar, India. He has more than a decade of teaching research experience. Dr. Pattnaik has published numbers of research papers in peer-reviewed international journals and conferences. His research areas are cloud computing, mobile computing, and brain-computer interface. He has authored many computer sciences books in the field of robotics, turing machine, cybersecurity, cloud computing, and mobile computing. Dr. Pattnaik is a Fellow of the Institution of Electronics and Telecommunication Engineers.

Contents

Contributors

Rishabh Arora
School of Computer Engineering, Kalinga Institute of Industrial Technology, Deemed to be University, Bhubaneswar, Odisha–751024, India

Sharmistha Banerjee
Kalinga Institute of Industrial Technology, Deemed to be University, India

Rabindra Narayana Behera
National Informatics Center (NIC), Bhubaneswar, Odisha, India

Abhinav Bhandari
Department of Computer Engineering, UCOE, Punjabi University, Patiala, Punjab, India

Poonam Biswal
Kalinga Institute of Industrial Technology, Deemed to be University, India

Poulami Bose
School of Computer Engineering, Kalinga Institute of Industrial Technology, Deemed to be University, Bhubaneswar, Odisha–751024, India

Balwinder Singh Brar
Department of Applied Sciences, Baba Farid College of Engineering and Technology, Bathinda, Punjab, India

Suchismita Das
Kalinga Institute of Industrial Technology, Deemed to be University, India

Sarita Dhal
School of Humanities, Kalinga Institute of Industrial Technology, Deemed to be University, Bhubaneswar, Odisha, India

Patel Dhruv
Kalinga Institute of Industrial Technology, Bhubaneswar, Odisha, India

Shreshtha Ghosh
School of Computer Engineering, Kalinga Institute of Industrial Technology, Deemed to be University, Bhubaneswar, Odisha–751024, India

Nishtha Jaiswal
Kalinga Institute of Industrial Technology, Deemed to be University, India

Pranjal Kumar
Kalinga Institute of Industrial Technology, Bhubaneswar, Odisha, India

Sushil Kumar
Department of Research, Innovation and Consultancy, I. K. Gujral P.T.U, Kapurthala, Punjab, India

Jibendu Kumar Mantri
Department of Computer Application, North Orissa University, Odisha, India

Deepanjali Mishra
School of Humanities, Kalinga Institute of Industrial Technology, Deemed to be University,
Bhubaneswar, Odisha, India, E-mail: deepanjalimishra2008@gmail.com

Nishikanta Mishra
School of Humanities, Kalinga Institute of Industrial Technology, Deemed to be University,
Bhubaneswar, Odisha, India

Rajalaxmi Mishra
College of IT and Management Education, Bhubaneswar, Odisha, India

Proshikshya Mukherjee
KIIT Deemed to be University, Bhubaneswar, Odisha, India

Soumya Mukherjee
Government College of Engineering and Ceramic Technology, Kolkata, West Bengal, India

Rajat Rajesh Narsapur
School of Computer Engineering, Kalinga Institute of Industrial Technology, Deemed to be
University, Bhubaneswar, Odisha–751024, India

Subham Naskar
Kalinga Institute of Industrial Technology, Bhubaneswar, Odisha, India

Swapnamoyee Palit
School of Humanities, Kalinga Institute of Industrial Technology, Deemed to be University,
Bhubaneswar, Odisha, India

Manaswinee Madhumita Panda
Department of Computer Science and Engineering, Chitkara University Institute of Engineering and
Technology, Chitkara University, Punjab, India, E-mail: mmpandacet@gmail.com

Surya Narayan Panda
Department of Computer Science and Engineering, Chitkara University Institute of Engineering and
Technology, Chitkara University, Punjab, India

Prasant Kumar Pattnaik
School of Computer Engineering, Department of Computer Science and Engineering,
KIIT University, Bhubaneswar, Odisha–751024, India

Ankit Pradhan
Kalinga Institute of Industrial Technology, Deemed to be University, India

Bhaswati Sahoo
School of Computer Engineering, Kalinga Institute of Industrial Technology,
Deemed to be University, Bhubaneswar, Odisha, India

Mangal Sain
Dongseo University, South Korea

Sasmita Rani Samanta
KIIT Deemed to be University, Bhubaneswar, Odisha, India

Manishankar Sannigrahi
School of Computer Engineering, Kalinga Institute of Industrial Technology,
Deemed to be University, Bhubaneswar, Odisha, India

Meet K. Shah
School of Computer Engineering, Kalinga Institute of Industrial Technology,
Deemed to be University, Bhubaneswar, Odisha–751024, India

Manish Snehi
Department of Computer Engineering, UCOE, Punjabi University, Patiala, Punjab, India

Sumesh Sood
Department of Research, Innovation and Consultancy, I. K. Gujral P.T.U, Kapurthala, Punjab, India

Prasanta Kumar Swain
Department of Computer Application, North Orissa University, Odisha, India

Sukanta Chandra Swain
KIIT Deemed to be University, Bhubaneswar, Odisha, India, E-mail: sukanta_swain@yahoo.com

Abbreviations

AHP	analytic hierarchy process
AI	artificial intelligence
ANFIS	adaptive neuro-fuzzy inference system
ANN	artificial neural network
AR	autoregressive
ARIMA	autoregressive integrated moving average
B2B	business-to-business
B2C	business-to-consumer
BDA	Bhubaneswar Development Authority
BPR	backpropagation with Bayesian regularization
BSCL	Bhubaneswar Smart City Limited
BTC	Bhubaneswar Town Center
C	culture
CBC	cell broad cost
CERT	computer emergency response team
CH_4	methane gas
CNY	Chinese Yuan
CPS	cyber-physical systems
CSCL	computer-supported collaborative learning
CVS	computer vision syndrome
DDoS	distributed denial of service
DE	differential evolution
DeitY	Department of Electronics and Information Technology
DHS	Department of Homeland Security
DLP	discrete logarithm problem
DNN	deep neural network
DOS	denial of service
E	environment
EC	economy
ECDLP	elliptic curve discrete logarithm problem
EDI	electronic information exchanges
EFT	electronic subsidizes move
EGCC	Energy Sector Government Coordinating Council

EMR	electro-magnetic-radiations
EPAN	European Public Administration Network
E-VCR	vehicle check report
FDRI	financial resolution and deposit insurance
FFNN	feed-forward neural network
FLANN	functional link artificial neural network
G2B	government-to-business
G2C	government-to-citizen
G2G	government-to-government
GA	genetic algorithms
G-AHP	grey-analytic hierarchy process
GBV	*gender-based violence*
GMC	Goa Medical College
GST	goods and services tax
HIS	hospital information systems
HMM	hidden Markov model
HRIDAY	Heritage Cities Development and Augmentation Yojana
ICTs	information and communication technologies
IDEMA	International Disk Drive Equipment and Materials Association
IIoT	industrial internet of things
IMF	International Monetary Fund
IoT	internet of things
IT	information technology
IVR	interactive voice response
LSTM	long short-term memory
M	management
MA	moving average
MAE	mean absolute error
MAPE	mean absolute percentage error
MLAT	mutual legal assistance treaty
MLP	multi-layer perceptron
MMP	mission mode project
MoUD	Ministry of Urban Development
MSDG	Mobile Seva Delivery Gateway
MSMEs	medium and small enterprises
NAT	network address translators
NCCOE	National Cybersecurity Center of Excellence

NCIIPC	National Critical Information Infrastructure Protection Centre
NeGP	National E-Governance Plan
NFC	near-field communication
NFV	network function virtualization
NICNET	National Satellite-Based Computer Network
NPPD	National Protection and Programs Directorate
NSDG	National Governance Service Delivery Gateway
ONF	open networking foundation
ONG SCC	oil and natural gas subsector coordinating council
PIS	positive ideal solution
PLC	professional learning community
POS	point of sales
PS	pound sterling
R	reliability
R&D	research and development
RAS	remote access servers
RB	resilient back
RBI	Reserve Bank of India
RCGPANN	recurrent Cartesian genetic programming evolved artificial neural network
RMSE	root mean square error
RS	Reed-Solomon
RTI	right to information
RTOs	Regional Transport Offices
SBM	Swach Bharat mission
SBP	standard backpropagation
SCG	scaled conjugate gradient
SCI	smart city initiatives
SCM	smart city mission
SCM	supply chain management
SMS	short message services
SPIC	Society for Promotion of IT in Chandigarh
SPSS	statistical package for social sciences
SPV	special purpose vehicle
SSDG	state e-governance service delivery gateway
SVR	support vector regression
T	technology

TAM	technology acceptance model
TCS	Tata Consultancy Services
TNSACS	Tamil Nadu State AIDS Control Society
TOD	transit-oriented development
TSA	Transportation Security Administration
TWD	Taiwan dollar
ULBs	urban local bodies
UNPAN	United Nation Public Administration Networks
USD	United States Dollar
USSD	unstructured supplementary service data
WAP	wireless application protocol

Preface

e-Governance refers to the use of the electronic mode of communication for the availability or various government services. The adoption of e-Services in the government sector built citizens' trust and reliability of the available services. The Ministry of Information and Technology (MeIT) has made e-Development of India as its medium. With the help of strategic e-Infrastructure creation to facilitate and promote e-Governance, electronics, and information-enabled services, the security of India's cyberspace, etc., India being a developing country, is competing with various developed countries and has drastically changed its modes of governance services from manual human force to an electronic mode of services. In this era of digitization, fast, reliable, and transparent services are always encouraged. All these e-Services are working in a cyberinfrastructure using internet services, electronic devices, and online transactions. Therefore, the infrastructure needs to be highly secured for better communication with full protection against any attacks. Cybersecurity strategies and policies help to maintain proper synchronization in delivering all the services to the citizens efficiently and securely.

Globally, proper implementation of e-Governance services with proper security and protection has a great impact on the global value in the economy. In the global value chain model, all the aspects of the economy are highly interconnected with the governance services, cyberinfrastructure as the communication mode, which affects the global economy. Therefore, these three are the key concepts for a better economy globally using secured e-Services for growth in the economy.

This book organized into 17 chapters:

In Chapter 1, an introduction to the need of cybersecurity in e-Governance system is discussed and focuses on cybersecurity concept in e-Governance, the e-Governance model, its issues, and the process of how to maintain security in e-Governance. Chapter 2 introduces the need for cybersecurity in the governance system and its role in the global value chain governance policy. Chapter 3 is a theoretical study on how women fall prey to social networking sites and remedial actions to be taken for preventing such measures. Chapter 4 aims at unfolding how technology adoption and sustainability have been a challenge now in the banking industry through direct personal interviews of 100 retail-banking customers based at Bhubaneswar. Descriptive statistics are used to analyze and interpret the collected data.

Chapter 5 emphasizes one-government security and trust in order to maintain purpose among stakeholders to consume, process, and exchange the information over the e-government systems. G-AHP (grey-analytic hierarchy process) method is used in this chapter to help policymakers conduct a comprehensive assessment of e-government security strategy. Chapter 6 explains the uses of various e-Governance policies in the healthcare sector and aims to increase awareness of ICT use in the health sector, to evoke interest in medical students in learning ICT and to save precious time of both patients and doctors. It encourages the use of e-Systems in order to provide quality healthcare services at the doorsteps from anywhere and anytime.

Chapter 7 describes the different types of vulnerabilities and attacks that are found in the oil and gas industry. These attacks are life-threatening as there is the involvement of various people at various levels of operations. A slight manipulation in the design can cause a serious hazard, which may put workers' lives at risk. Chapter 8 describes an encryption scheme using ECC and Arnold's transformation algorithms for message communication through the network. Message encryption plays a vital role in communication. A small change in message can showcase different meaning of the communication.

Chapter 9 explains a software-defined networking (SDN) that has the ability to easily scale, extend, and dynamically add network functions to make the networking secured. In this chapter, the importance is given to FoG Computing, which acts as an additional layer for reducing latency between perception layer and cloud as well as to address the security concerns, such as detecting DDoS attacks. Chapter 10 gives a comparative evaluation of free/open source and proprietary software through a survey conducted for Indian users. FOSS uses users' response and analyzes the information. Many payment schemes are insecure and vulnerable. Therefore, FOSS analysis can be proven a better tool for such gateways.

Chapter 11 focuses on the various cyber threats that are being witnessed in this cyber world. The field of science that studies the various cyber-attacks and the consequences is referred to as cyber forensics. In this chapter, an overview of various cyber threats is studied, and the phases of digital forensics are discussed. Chapter 12 describes the development of information communication systems including social issues. The author focuses on the overall economic development of India. Some issues include use of artificial intelligence (AI) as a part of transmission of information, controlling the greenhouse effect, industrial and agricultural development with the use of technology, and many more.

Chapter 13 gives an analysis of the role of digital technology in collaborative technology. The collaborative method provides versatile productivity in a creative and atmosphere of mass accountability that gives rise to more benefit. The positive impact of collaborative learning in various studies is appraised by the improvements of personal attributes ignoring irrelevant failures and drawbacks in a leading-edge ambiance. Individual learning in an isolated atmosphere may demonstrate a paucity of creative, cognitive, moral reasoning, on the contrary, be supplemented by a variety of cultural and ethnic backgrounds, cooperation supplanting competition. Therefore, this chapter proposes a critical analysis of the concept of collaborative methodology, challenges subjugated in this domain and how they may be overcome. Chapter 14 provides a survey report on the forecasting of the currency exchange rate methodologies. Currency exchange plays a vital role in the economy. The forecasting methodologies would give immense knowledge on the various exchange rate methodologies. In this literature survey, many techniques are discussed, of which some are as good as random walk models or slightly worse than the projected model.

Chapter 15 illustrates the issues and challenges of e-Governance and m-Governance in India. e-Governance and m-Governance are two major pillars of digital India. Most of the governance systems are working under these schemes for providing better service facilities to the citizens of India. Therefore, in this chapter, different factors like rules and regulations, the establishment of information security system, optimization of business process, and evaluation of e-Governance are discussed, which are responsible in order to have a reliable and good m-Governance and e-Governance services in India. Chapter 16 presents a case study of smart cities. It describes the various possible solutions that are implemented in the smart cities for good governance and better service to the citizens. The G2C (government-to-citizen) and C2C (citizens to citizens) connectivity made possible by the wave of technology has further strengthened the democracy with every issue getting overall public vision and opinion. The impact in its wide bloom can only be known after some more years in terms of its overall influence on total income, job generation, economic growth and development, etc., as the multiplier process would take its lag. This chapter discusses the existing issues and apprehensions related to the smart city and its technological disruption, which confront its economic contributions.

Chapter 17 provides an overview of the blockchain application focusing on storing information about the digital signatures in distributed networks. It also gives the taxonomy and architecture of blockchain and aims to conduct the survey, identify, analyze, and organize the literature on blockchain in supply chain management (SCM) and industrial applications. It also inspects the proof of stakesible enhancements that blockchain would provide, including major disruptions and challenges that arise because of its adoption.

—Editors

CHAPTER 1

Introduction to Cybersecurity in e-Governance Systems

MANISHANKAR SANNIGRAHI,[1] BHASWATI SAHOO,[1] and
RABINDRA NARAYANA BEHERA[2]

[1]*School of Computer Engineering, Kalinga Institute of Industrial
Technology, Deemed to be University, Bhubaneswar, Odisha, India*

[2]*National Informatics Center (NIC), Bhubaneswar, Odisha, India*

ABSTRACT

The operations of e-Governance have been increased significantly because of the citizen's demand of cost-effective and timely services. The growing span of e-Governance and its impact over the community proves that it is more than a single system. The government has made an outgrowth effort to improve their relation with citizens through e-Governance. Electronic transactions shall have the equivalent legal value such as other process of communication like written form. Information security practices such as policies, procedures should be placed and security technologies must be utilized to protect e-Governance system from attacks, threats. There are two main fundamental factors one is required level of integrity and authentication must be provided to the public key infrastructure. Second one, the security of e-Governance can be maintained by making people realize the significance of the issue. An awareness program should be conducted for the people, so that they can recognize the threats, issues of security and learn how they can resolve them. This chapter consists of the cybersecurity concept in e-Governance, e-Governance model, its issues, and the process of how to maintain security in e-Governance.

1.1   INTRODUCTION

The e-Revolution has cleaned the business by creating e-business and e-commerce. e-Government embraces the e-revolution because of the same reason. Customer of the service has increased significantly in last few years which bring the business to e-business. Because of this new method the businesses have to adapt new operating procedure to capitalize and the distribution channels like business-to-consumer (B2C) and business-to-business (B2B) have also altered. e-Commerce has strengthened the communication channels because of which the internal and external operations have nourished. e-Commerce model is totally different from the normal business model.

Government has made several efforts to improve relation between the government and its citizens. e-Governance is one of those efforts, with its transparency and openness it brings the government, and the citizens more close. e-Governance has wide social circle that ensures the representation of democracy, the capability of adapting in the changing environment by the knowledge of continuous generation applications and techniques give the economic and competitive advantage. e-Governance faces many challenges. Adapting new delivery model in the case of electronic requires modification in policies and procedures. There are some categories in e-government like government-to-government (G2G), government-to-citizen (G2C), and government-to-business (G2B) [1].

The required modifications create new opportunities and problems for the government agencies which are moving towards the new media. Record security has long term proven solutions, these solutions arises new issues. Access is limited to secure the physical storage, but in electric records it is very difficult to control the access. It will cost both time and money to access a paper. The digital world makes multiple remote accesses for electronic records and databases. It saves both time and money, but this efficiency comes with a new question how the access can be controlled and secured. There is a common format which has been used to record paper for 100 years, but the electronic records have its own issues though it is a few decades old.

In the 21st century, threat to cybersecurity is the biggest existed challenge. Threats are coming from different sources and involve in disruptive activities against individuals, business, government, etc. The effects of these threats bring significant risks to public safety, the national security, and the international community. It is easy to conceal the malicious use of information. The identity and motivation of a perpetrator are very difficult to understand; we can understand it by analyzing the circumstantial evidence, from the consequences of the attack and the target. Perpetrators can attack from anywhere in the world. The motives can simply be the demonstration of technical prowess, money, revenge, etc. Criminals and hackers have used and created many malicious tools and methodologies to operate. The growing sophistication in criminal activity has increased the harmful actions and makes it harder for the authorities to prevent it.

1.2 THE CONCEPT OF CYBERSECURITY

The increasing concerns about protecting ICT systems from cyberattacks have been expressed by the policymakers and experts in past years. Unauthorized persons try to access the ICT systems for various unlawful actions. The number and severity of cyberattacks will be increased in next several years. Cybersecurity is an attempt of securing the ICT systems and its data from attackers. It refers to three main things:

1. It is the set of measures and activities to secure computers, networks, hardware, information, and software devices from disruption, threats, and attacks. It also meant to protect communication, software data, different cyberspace elements;
2. The quality of protecting cyberspace and its elements from threats; and
3. The attempt of the board is aimed to improve and implement the activities of protecting cyber world.

Cybersecurity and information security are not conceptually identical. Information security is described as defending systems and its information from discloser, modification, unauthorized access, modification. There are three goals of information security:

1. **Integrity:** It means protecting information from unauthorized modification and destruction to ensure information authenticity and nonrepudiation;
2. **Confidentiality:** It means preventing unauthorized access and information discloser, includes protecting proprietary information and privacy; and

3. **Availability:** It means reliable access of information and timely use of that information.

Cybersecurity is combined with other concepts like privacy, intelligence gathering, sharing of information, surveillance, etc. Privacy is the concept of protecting someone's private information from other individuals or agencies. Cybersecurity can be a useful tool to protect privacy in digital environment. Information which is used to help cybersecurity efforts may contain information which can be considered as private. Cybersecurity is used to protect people from unwanted surveillance and intelligence gathering. Gathering of intelligence can help cybersecurity to track potential cyberattack sources. If surveillance is use to monitor the flow of information from or within a system it can be useful for cybersecurity [12].

1.3 CYBERSECURITY ISSUES

Computer security history has shown regression. At the early stage computer systems are highly secured, but if we compared it with today's systems it performs very less in availability. Today's computers has many functionalities, features, availability of data has been increased significantly. This increase availability of data has some issues of data integrity and confidentiality. The principle of developing software and system is added more features first, think about security later. The software industry has started to put their attention to the increased security issues in last few years.

The internet was designed on the basic platform of trust and shared access but with less security. Most of the protocols and policies that provide little security to the users are based on trust; at the early stage of internet development this model had some sense because the transferred data is only valuable to owners not others. Today internet is worked as a key tool of transferring data between people, businesses, government departments, other entities. People have shared their bank details, private information, etc., through internet; this information is valuable to others especially to the criminals and cyber attackers. They use many techniques to obtain the information like phishing attacks, identity thefts [10].

In the current environment information has significant value and to protect it the software developers and designers has the responsibility to design systems with a higher level of security to provide the users an appropriate CIA level [7]. This is the reason of having many complexities in e-government and e-commerce. It is very difficult to prove someone's identity in social media or to a software program. Government documents are very vital for our society. The documentation acts of the government are the foundational elements to support someone's identity. People are using driving license, taxpayer identification number and other documents to prove their identity and to get authorization for many opportunities. Criminals are targeting for these documents because of their importance. To maintain appropriate cybersecurity we need to combine technical and managerial departments. Management can set the risk tolerance level and decide the requirements of security. Technical team after analyzing the requirements can design and deploy appropriate safeguard for securing the system and data. There should be an auditing and testing system to see the solution correctness and to state further requirements [11].

Security failures are considered to be high profile incident. Millions of records have been exposed because of identity theft in Bank of America, Card systems, DSW [2]. These events are bad for business and PR can destroy company and its reputation. There are some marketplace remedies for companies which are not performing according to the standards of the community. However, for the government there are no such remedies available.

1.4 e-GOVERNMENT MODEL

The basic operations of government are different from the operations of business that's why e-government is not an extension of e-Business. Government operations have different entities starting from differing constituents, independent agencies operating differently to each other and there is a limitation on the ability of raising capital. The biggest problem of e-government is its multiple no of departments and agencies. These agencies are also being the customer like local firms and citizens. There is a priority level in information sharing, information shared between city water department and police department can have higher level of priority then information shared between the citizens and police department. These entities have important role towards the response of a cyber-security incident. Interactions between these different entities can be seen as communication channel in terms of content and trust. Managing the relationship between channels is an exponentially growing challenge as the no of channel is increasing [1].

The hierarchical order is maintained in government agencies but the interagency communication is not clearly manifested. In business, the interdepartmental problems are solved by the quick action of the higher up; the response is focused on the mission. Government has o many relationships and departments to find a common solution for coordination issues. Government agencies are not focused in capital development; it hinders their response to the demands and new opportunities of customer. Government serves for the benefit of the society. Commercial firms benefit their shareholders, government benefits only the community they work for. This connection with community is important for the government operation. Commercial firms are the source of a specific product or service, but government is the source of various services to the community. This view can be positive or negative; people always want the best thing with lower cost even when the cost is not enough for the service [3].

e-Government can be automated but it will be a costly. Software and hardware need resources for that. The features related to security will be the first one to be cut because of low budget. Providing the services to the end users is the first priority and security issues will be considered after an incident. The national agencies have given attention to the authentication methodology as a common hardship area. New solutions have given the e-Government insight to its unique challenges. Significant information can be found by using national level resources to handle technical issues related to security. These resources are unfortunately used as a stovepipe towards other technological aspects. Furthermore, government entities do not have the capital resources like commercial entities to support their functionality. Government has limited financial resources, but the important thing is how to use them effectively. It department must come with a plan to maximize the effects of the application with finite resources in the area of information security. The department should use some scale of measurement to measure the effectiveness.

1.5 CYBERSECURITY IN E-GOVERNANCE

1.5.1 CLASSIFICATION

The meaning of managing e-Governance is to manage large portfolios of different responsibilities in a reasonable manner with all subjects consist of e-Governance usage and implementation. All the users of the system need to know about the system to develop a good e-government system. e-Governance is classified into four main categories the government, citizens, business, and employee. All of them are inter linked (Figure 1.1).

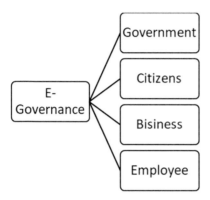

FIGURE 1.1 Classification of e-governance.

Various models are available for each sector to implement e-Governance. The actions of defending systems and data regarding networks, data storages, application software's, etc., by using proper technology is known as cybersecurity. Cyber defense is related to special activity which is linked to particular facets and organization. The factor which distinguishes cyber defense and security is the assets to protect and the mechanism used to protect them. The defensive actions which are taken because of the activities originate from hostile actors motivated by political or financial agenda have the effect on national security and economic stability of a country. Technologies which have the capability of incident response, real-time protection are deployed to cyber defense [8]. This lead the way of creating ICT systems for the government.

The key element of cybersecurity in e-governance is the partnership between public and private. These partnerships can deal with coordination problems. They are capable of increasing cooperation and information exchange. The engagement can take different forms and able to address training, awareness, recovery operation, etc. The supply chain of Information and Communication Technology (ICT) can be affected in a certain way that can affect secure, reliable, and normal use of information through much application. This is becoming a concern for the states around the global. Disruptive activities can grow dangerous and complex in cyberspace.

Collaboration between states, civil society, and private sector is important to confront the challenges of the future, and the effective measures to upgrade cybersecurity require international cooperation.

1.5.2 APPLICABILITY

It is required to investigate both the critical success and the challenges of the e-Governance. To assure security in cyberspace we need to be careful and attentive towards the creation of the system. In the process it is important to engage right people with suitable behavior, ethics, and awareness, also the use of appropriate technology. The cyberspace and information technology (IT) has transnational characteristics. A series of process is required to create a secure cyber ecosystem. Collaborative, cooperative, and direct actions which are taken by the state and beyond to establish a secure cyber ecosystem are following:

1. To determine and implement suitable response against a threat it is essential to create awareness in the infrastructure of Information and Communication Technology (ICT) [9].
2. Creating a good legal environment to support secure and safe cyberspace, sufficient confidence and trust in electronic transactions, strengthen the capabilities of law and enforcement so that reasonable action can be taken by stakeholders.
3. Protect the gateway and networks of information and critical communication infrastructure.
4. 24×7 mechanisms should be in place to emergency response, crisis management through protective, preventive, predictive, recovery actions.
5. Should focus on developing indigenous technology and security techniques through solution-oriented and frontier technology research, pilot development, etc.
6. Should include cybersecurity in our culture for appropriate user action and behavior.
7. Should put effective cybercrime prosecution act in every ICT environment and also have some cybercrime prevention procedures.
8. Protect the data while it is in transit, in storage, in process and also protect personal information of the user to create a trustworthy environment.

It is the responsibility of the government agencies to pay attention in the development of proper policies and guidelines of information security and encourage people to use appropriate application and technology in the organization. There is a need of creating a framework of information security. The aim of the framework is to combine efforts from all applicable groups to protect critical information.

1.6 CONCLUSION

In the global economy, e-Governance has its own significant place. World Bank and UNO has provided support to the initiatives of e-Government. The actions of securing information systems and information are done at different level [4]. According to many researchers Government should be transparent while functioning and should introduce legislation if needed. There is a need of legislative changes in various e-Governance entities like electronic archiving, signatures, data matching, computer crime, intellectual property rights, etc., [5]. The Government has the responsibility to bring strong legislation and security in cyberspace to reduce the misuse of internet. Besides the Government actions the home users, small users, service providers also understand their responsibility to play active part in the security of cyberspace. It is necessary to fight against cybercrime if we want to make technology trustworthy and beneficial for the public.

1.7 FUTURE WORK

The next step of the work will be focused on the communities of the regional and state level [6]. There is a limitation between the links of direct communication in the top level. However, the greatest challenge of all is the dependencies in the inter-agency communication at the regional level. The interaction between the highest level and local level government agencies is very limited and the only way is to go through the channel of higher agencies. We should add electronic access challenges in ales mature and new environment; this way we get to know the existed gaps, what we can ignore, and what we need to address first.

After looking at the issues we can make many government agencies and many communities aware about new opportunities and issues. We can make them aware about cyber threats and cybersecurity policies and procedures. It is the infancy level of e-Government and other initiatives. There will be a growth and learning period, after that lessons will be learned and a better solution will be given and implemented.

KEYWORDS

- **business-to-business**
- **business-to-consumer**
- **cybersecurity**
- **e-governance**
- **government-to-business**
- **information and communication technology (ICT)**

REFERENCES

1. Carter, L., & Belanger, F., (2005). The utilization of e-government services: Citizen trust, innovation, and acceptance factors. *Journal of Information Systems, 15*(1), 5–25.
2. Acohido, B., & Swartz, J., (2005). *ID Thieves Search Ultimate Pot of Gold Databases.* In USA Today.
3. Alford, J., (2002). Defining the client in the public sector: A social-exchange perspective. *Public Administration Review, 62*(3), 337–346.
4. Kumar, D., & Panchanatham, N., (2014b). Strategies for effective e-governance management. *International Journal on Global Business Management and Research, 3*(1).
5. Kumar, D., & Panchanatham, N., (2015a). A study on cyber law in promoting e-governance. *AE International Journal of Multidisciplinary Research.*
6. White, G., Dietrich, G., & Goles, T., (2004). Cybersecurity exercises: Testing an organization's ability to prevent, detect, and respond to cybersecurity events. In: *Proceedings of the 37th Hawaii International Conference on Systems Science.* Kona, HI.
7. A Report from CISCO, (2010). *Cybersecurity: Everyone's Responsibility.*
8. Rathore, S., & Dubey, (1998). *Barriers to Information and Communication.* Technologies Encountered by Women Sponsored by the Commonwealth of Learning and the British Council, India.
9. Aibara, H. D., (2017). *Introduction Information and Communications Technology (ICT).* https://shodhgangotri.inflibnet.ac.in/bitstream/123456789/4132/./02_introduction.pdf (accessed on 3 November 2020).

10. A report available at: http://www.sse.gov.on.ca/mcs/en/pages/identity_theft.aspx (accessed on 3 November 2020).

11. Lipson, (2002). *Tracking and Tracing Cyber-Attacks: Technical Challenges and Global Policy Issues*. Carnegie Mellon Software Engineering Institute, Pittsburgh.

12. Goutam, R. K. (2015). *Importance of Cybersecurity*. Semantic Scholar https://pdfs. semanticscholar.org/5cfb/7a5bd2e6c181e8a69ebd49b1dadb795f493b.pdf (accessed on 3 November 2020).

CHAPTER 2

Digitization and Its Impact

SUBHAM NASKAR,[1] PATEL DHRUV,[1] PRANJAL KUMAR,[1] and
SOUMYA MUKHERJEE[2]

[1]*Kalinga Institute of Industrial Technology, Bhubaneswar, Odisha, India*

[2]*Government College of Engineering and Ceramic Technology, Kolkata,
West Bengal, India*

2.1 INTRODUCTION

The world has undergone massive changes since the ushering of the
digital age. Digital devices and technologies have assisted our lives in
every possible way. Digitizations have challenged the convention in many
spheres. Traditional paths have been enhanced or often replaced by digital
means. We have tried to document three important aspects of such a change
in this chapter. Traditional services required human interaction, many of
which have been replaced by e-services. Cyberwarfare has the power to
cripple security, infrastructure, and economy without using any weaponry.
We have seen small-scale cyber-attacks again countries in recent times.
Cybersecurity has thus the necessary protection against possible cyberwarfare. Many countries are increasingly bolstering their assets in cyber
space. Ever-changing political landscape has seen certain countries of
the world to coagulate under major groups. Value chain policy is very
important for cooperation between adjacent countries for economic and
regional growth. We look how such policies have been evolving over time.

2.2 ADVANCEMENT IN E-SERVICES

e-Services (electronic services) are the implementation of information and
communication technologies (ICTs) to provide services to people or some

particular organization. e-Services constitute of three main components namely:

1. Service provider;
2. Service receiver; and
3. Channels of service delivery.

For instance, in public e-services, agencies (government) are service providers and receivers, businesses are the service receivers. The method of administration conveyance is the third necessity of digital administration. Web is the principle conveyer of e-administration conveyer while there are other exemplary ways prevailing (for example, phone, call focus, open booth, cell phone, TV) are likewise considered [1].

The term 'e-service' has many applications and can be found in many parts of e-trade. Two main areas of e-services application are as follows:

- **e-Business:** e-Services or e-commerce provided by businesses (private sector); and
- **e-Government:** e-Services provided to citizens or business by the government of the country.

2.2.1 E-BUSINESS

e-Business refers to the involvement of electronic innovation in business functionalities. It administers the utilization of PCs and advanced gadgets for conducting activities like Internet for assistance and web-based businesses. The advancement of organizations toward expanding e-business capacities dates from the across the board utilization of PCs during 1980s and upscale improvement of the business Internet during the 1990s. This advancement is continuous as e-services progress to versatile innovation [2].

2.2.1.1 FACTORS CONTRIBUTING TO THE RISE OF E-BUSINESS

2.2.1.1.1 Information Storage

The development of e-business has aided organizations to upgrade from writing down information on paper to advanced stockpiling of relatively much more information on servers. Electronic capacity has helped organizations to increase the volume of information stockpiling in lesser time and using less effort. The work that was previously used to be done physically with multiple involvements became quicker through broad utilizing spreadsheets and committed digital approaches. Organizations store information on client cooperation and are capable of processing the information rapidly. Future transformative patterns tend toward storing information away in the cloud [3].

2.2.1.1.2   Internet

The Internet has become a key e-business innovation. Advancing from a military and instructive system to a business Internet in the mid-1990s, the premise of sites and web-based business was set with the improvement of the program Netscape in 1994, and the establishing of Amazon and eBay in 1995 [4]. The biggest organizations were aided by the approach of communication through the Internet, and many offered merchandise and ventures for purchase on their sites. As the Internet kept a steep curve in advancing, even ventures that started working independently created their own websites. With cell phones creating more Internet traffic, sites have become progressively versatile option, with devoted portable applications permitting cell phones and tablets to get the data [3].

2.2.1.1.3   Communication

The new advances of electronics have changed correspondences in business. e-Business rose from pre-electronic paper mail, wire transmissions, and print to email, messaging, and fax, with the last now very nearly outdated nature. On the phone side, long separation calls turned out to be significantly more affordable and Voice over Internet Protocol calls can be totally allowed to any Internet association. New advances like Skype and Google Hangouts make free video conferencing conceivable. Organizations utilize these new e-business advancements to impart all the more seriously with their representatives and clients, and even to save money on movement costs for face-to-face gatherings [4].

2.2.2   MAJOR YEARS IN THE HISTORY OF DEVELOPMENT OF E-COMMERCE [6]

- **1991:** Creation of the World Wide Web.
- **1994:** Navigator launched by Netscape, and Pizza Hut offered online ordering.
- **1995:** First sale was done by eBay and Amazon.
- **1996:** Over 40 million people received Internet access, and online sales surpassed 1 billion dollars in the year.
- **1998:** PayPal was formed, changing the way people made online payments.
- **2000:** Revenue from U.S. online shopping generated over $25 billion.
- **2001:** During the holiday season around 70% of Internet users make a purchase online.
- **2003:** iTunes was launched by Apple, which was the first digital music store.
- **2005:** The term "Cyber Monday" was coined and gained one of the biggest online shopping days of the year.
- **2006:** Facebook begins advertisements sell.
- **2008:** For the first time products were bought online using mobile phones.
- **2012:** B2C sales surpass $1 trillion online.

2.2.3   E-COMMERCE OF EVOLUTION

The advancement of online business was a blend of advancement along with innovation. In spite of the fact that Internet which was considered as one of the primary reasons of development showed up in the late 1960s, e-Commerce currently took off with the appearance of the World Wide Web and programs during the 1990s. Gone back to the mid-1970s with advancements as if electronic subsidizes move (EFT)-assets can be steered electronically starting with one association then onto the next e-Commerce made a presentation. Electronic information exchange (EDI) was utilized to move routine reports through advanced implies, which extended their utilization from budgetary exchanges to different sorts of exchange preparing. Between the authoritative frameworks (IOS) permitted flowing of a stream of data between associations so as to arrive at an ideal supply chain management (SCM) framework, which empowers supported improvement [5].

Years showing important historical events which helped in the evolution of e-commerce [6]:

- **1984:** EDI, or electronic information trade, was organized through ASC X12. This guaranteed that the associations would have the choice to complete trades with one another reliably.
- **1992:** CompuServe offers online retail products to its customers. This gives people the primary chance to buy things from their PC.
- **1994:** Netscape appeared. Giving customers a clear program to surf the Internet and a safe online trade by means of advancement called Secure Sockets Layer.
- **1995:** Two of the best names in online business are impelled: Amazon.com and eBay.com.
- **1999:** Retail spending over the Internet comes to $20 billion, as per Business.com.
- **2005:** The improvement of the web has a remarkable centrality on the advancement of online business. Since the web is unobtrusive now on account of specific advances, it allowed the incorporation general people into the method.

In 2019, retail web based business deals added up to 3.53 trillion US dollars worldwide additionally e-retail incomes are required to develop to 6.54 trillion US dollars in 2022 (Figure 2.1) [7].

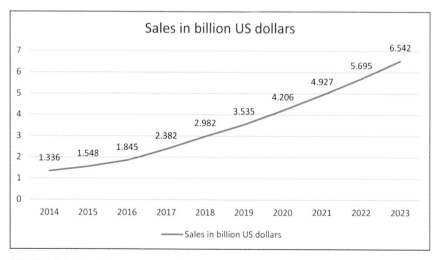

FIGURE 2.1 Retail ecommerce sales from 2014 to 2013.

2.2.4   E-GOVERNMENT

2.2.4.1   EVOLUTION OF E-GOVERNANCE

With the introduction of the World Wide Web (WWW), introduction of IT by governments developed and expanded in the nineties. The innovation has progressed a considerable amount and subsequently offering of IT by governments has expanded. Boundless availability of innovation has opened a very enormous extension in this segment also residents presently expect more data and services than ever before [7].

To catalyze the improvement of e-service laws and advancements in creating nations there has been expanding inclusion of worldwide organizations for advancement. While the highlight has been basically computerization, state governments have also attempted to use ICT instruments into the accessibility of information arranging information, setting up structures for taking care of information and passing on administration. At a little scale level, this has stretched out from IT digitalization in solitary workplaces to interfacing every one of the workplaces, racks of papers containing information to electronic information recording, access to training and employment-related things, open grievance entries, administration transport for high volume routine trades, for instance, portions of bills, and so forth.

2.2.4.2   TRENDS IN TRANSACTIONAL ONLINE SERVICES

Pretty much every nation is offering some e-Services yet not all them are offering on the web Transactional services. Be that as it may, nations offering this service have expanded from 18% to 47% in all service classes. Paying for utilities (140 nations), income tax returns (139 nations); what's more, enrolling new organizations (126 nations) are the three most normally utilized online services in 2018.

2.2.4.3   DISTRIBUTION OF ONLINE SERVICES BY SECTOR

The Internet, cellular gadgets and different agencies contributing to gather, preserve, examine, and covey data carefully are currently being embraced by different government segments. Nations offering services through various mediums like applications, SMS, Email, and so on in various parts have expanded (Figures 2.2 and 2.3).

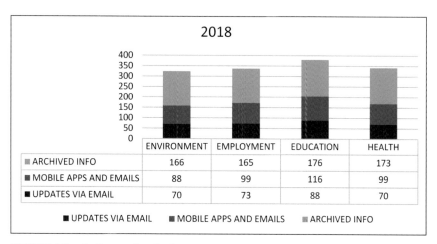

FIGURE 2.2 Online services in the year 2018.

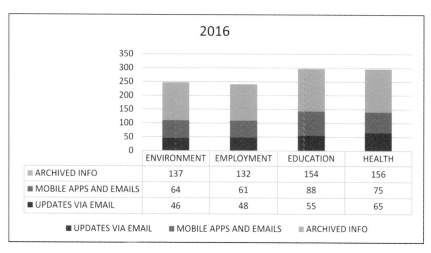

FIGURE 2.3 Online services in the year 2016.

e-Services offered through portable Apps are becoming quickest, at 52%, in parts like education, work, and environmental sectors. In the business segment services by means of email and RSS have expanded the most, at 62%, trailed by environmental division, at 38%. In any case, less nations offer downloadable structures in the environmental sector in 2018 contrasted with 2016.

2.2.5   TARGETED SERVICES FOR VULNERABLE GROUPS

A constructive pattern recorded is showing that numerous nations are giving on the web services to poor people; in fact the number has almost tripled while those giving web service access to the young, ladies, transients, exiles, people with inabilities have doubled.

2.2.6   KEY DIMENSIONS OF GOVERNANCE FOR SUSTAINABLE DEVELOPMENT [7]

1. Accountability;
2. Effectiveness;
3. Inclusiveness;
4. Openness;
5. Transparency.

Mostly, the online services offered by the countries having Very High EGDI level cover more aspects of these principles than other countries.

A decent sign towards following these standards is the arrangement of open stages by governments for e-acquisition and bidding process. Open declarations about e-acquirement procedures and offering results, just as online instruments to screen and assess e-obtainment agreements. In 2018, a huge increment from 40 to 59% of nations offering these rather than a similar arrangement of services in 2016.

The number of countries which offered e-procurement tools increased drastically from 2016 to 2018. Use of online announcement tools increased by 39%. Usage of e-procurement platforms increased by 32%, online bidding procedures increased 29%. Monitoring and evaluation of contracts online were done by 51% more countries in 2018 than done in 2016.

Similarly, more countries used the online platform for sharing information about government vacancies and employment opportunities, henceforth the transparency in the recruitment process increased and at the same time greater employee participation is also encouraged (Figure 2.4).

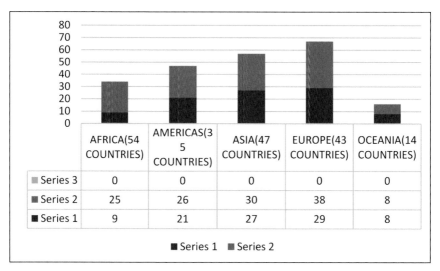

	AFRICA(54 COUNTRIES)	AMERICAS(35 COUNTRIES)	ASIA(47 COUNTRIES)	EUROPE(43 COUNTRIES)	OCEANIA(14 COUNTRIES)
■ Series 3	0	0	0	0	0
■ Series 2	25	26	30	38	8
■ Series 1	9	21	27	29	8

■ Series 1 ■ Series 2

FIGURE 2.4 2016 v/s 2018 online government vacancies.

2.2.7 GLOBAL DISPARITIES IN E-GOVERNMENT SERVICES

e-Government portals features can be divided in three categories based on the levels of the features which they incorporate:

1. **Basic:** Mainly focused on finding the portal, making popular and basic searches available, updating them regularly.
2. **Advanced:** Offering help, providing users with questions or FAQs that were frequently queried, introduction of feedback options for estimating public experience, provided links to for direct shop options, allowed users to access social media, and focused on making web accessible from any device.
3. **Very Advanced:** Allowed searching, increased availability of tutorials for helping the public to understand better about the system, help-desk optionality for remote aid from experts, also creating a facility that report unethical or unlawful and disturbing behavior.

Basic services are offered by almost all the nations. Low-income nations linger behind in implementing advanced and very advanced features.

2.3 ADVANCEMENT IN CYBERSECURITY

2.3.1 PREMISE OF CYBERSECURITY POLICY REQUIREMENT

Cybersecurity, in the broader sense of the term, involves preventive measures, pre-emptive detection, minimize the damage occurring form such attacks and give adequate response to the attacks in the cyberspace. Such cyber-attacks can affect an individual, corporate organizations, various social communities, and even worse at national level. Cyber-attacks are mainly unauthorized or often malicious attempts to control a computer system and create harm to more individual entities and cause damage in the different fronts including financial loss, identity theft, privacy invasion, and abuse, damage to reputation and safety.

With an exponential growth of the digital world and introduction of various interfacing and communication devices to the public, huge volumes of information including user data are being generated daily. These are vulnerable to malicious attacks, data abuse and misuse which can compromise public safety and national security. As reported by Statista [8] 157 information breaks were accounted for in the U.S. which uncovered 66.9 million records in 2005. In 2014, around 783 information ruptures uncovered at around 85.61 million complete records which indicated the addition of 500% in the augmentation of cyber threats. That augmentation example multiplied in three years with 1,579 detailed ruptures in 2017.

Development and deployment of cybersecurity policies are thus very important for a nation to safeguard its interest against unknown foes, various types of malicious actors, and create a secure and resilient cyber-space ecosystem where citizens, businesses, and the government itself can operate safely, without unknown actors intervening in one's privacy.

2.3.2 CYBERATTACK AND CYBERSECURITY

One of the key areas of interest in cybersecurity is not only to help in maintaining software integrity but also to prevent the data from being exploited by a third-party actor. A security breach not only adds up to heavy financial loses for a company but might also make personal and private data more vulnerable. Such infiltrations are termed as 'threats' which are of various types mostly developed by experts to get access to sensitive information of the victim with criminal and unethical intensions, human error and system fault might also result in the software integrity failure. Depending on different situations such threats have been categorized into: computer viruses, malicious security software, adware, and spyware, worm, denial-of-service (DOS), and distributed-denial-of-service attacks, SQL injection attacks, and phishing.

With new technologies being implemented every day to make tasks less time consuming, certain vulnerabilities are unknowingly kept unpatched which allow such malicious attacks. As concluded by Cybersecurity Almanac 2019 by Cybersecurity Ventures about 70% of cryptocurrencies are to be used for illegal activities which are up by 20% of all cryptocurrencies used in 2017 [9]. According to the Ninth Annual Cost of Cybercrime global study by Accenture security breaches rose about 11% over the last year and 67% over the last five years (Table 2.1) [10].

TABLE 2.1   Analysis of Attacks Caused by Ninth Annual Cost of Cybercrime Study

Type	Percentage Increase	Estimated Damage in 2017	Estimated Damage in 2019
Malware	+11	$2364k	$2613k
Denial of service	+10	$1565k	$1721k
Web-based attacks	+13	$2014k	$1396k
Malicious code	+15	$1282k	$1396k
Ransomware	+21	$532k	$645k
Malicious insider	+15	$1418k	$1621k
Botnets	+12	$380k	$390k

2.3.3   RECENT ADVANCEMENTS IN CYBERSECURITY POLICIES IN SOME COUNTRIES AROUND THE WORLD

The most recent cybersecurity policy of the United States of America was introduced by President Trump in 2018 [11]. While mostly staying in accordance to its course for its past initiatives, the new policy has been offensive in its outlook to counter threats from possible foes like China, Russia, Iran, and North Korea, and other non-state actors like terrorist organizations, criminal networks, and strategic adversaries. America's cybersecurity policy aims to publicize and mortify the cybercriminals and the countries backing them [11].

The offensive stance of the government is backed by the technological prowess of the Department of Defense, the NSA and other military branches of the US government. Thus, the shift in policy implies more proactive and offensive stance during any possible kind of cyberwar. This can turn out to be a double-edged sword due to the fact that any such retaliation can provide costly in a number of ways, while on the other hand can thwart a possible large scale cyberattack on the country's infrastructure.

Cyberattacks may mask their actual location using sophisticated masking software and pose themselves to originate from any third party neutral source or a non-hostile country. This may complicate international relations with that country and its allies. In order to avoid such cases, extremely careful steps are to be taken by the acting agency on behalf of the government.

The new cyber policy also allows NSA to make offensive moves against underground cyber black markets which deal with sensitive and often compromised data, and take down infected computers and network devices that form a botnet.

China has been laying its basis for cybersecurity laws since early 2003, with the "National Coordinating Small Group for Cyber and Information Security" (全国网络与信息安全协调小组) leading the charge. China introduced its first coherent cybersecurity policy in 2014 [12]. There was inconsistent and isolated cyber laws prior to this time. The cybersecurity affair was controlled by the State Council Internet Information Office on behalf of the government. The Chinese government pitches for locally coded programs and its use, despite the fact that western products are considered to be more secure and offer better performance. The main reason for this push is the belief that foreign products are a threat to national security. China's President Xi Jinping said "No national security without cybersecurity "to state-sponsored news agency Xinhua in April 2014 [13]. This clearly marks an encouraging and protectionist measure taken by the Government on its policy on cybersecurity.

In India, cybersecurity laws and policy were first introduced in 2013, in the light of the reports published by media outlets about NSA's active surveillance on India's internal politics, strategic, and commercial interests [14]. An updated policy is due for January 2020. Department of electronics and information technology (DeitY) under Ministry of Communication and Information technology oversees the implementation of the policy. National Critical Information Infrastructure Protection

Centre (NCIIPC) has been established to deal with cyber threats via a national and sectoral 24 × 7 mechanism. Computer emergency response team (CERT-In) has been built to serve as an umbrella organization and nodal agency to co-ordinate with sectoral CERTs during events of threats to the Information and communication technology (ICT) infrastructure, and serve effective response, resolution, and crisis management through appropriate threat resolution techniques. The policy calls for successful open and private organization collaborations and community-oriented commitment through specialized and operational participation. Promotion of research and development (R&D), and collaboration with "industry and academia" has been also emphasized in this policy. It also outlines the development of human resource through education and training programs and setting up cybersecurity training infrastructure with the aid of public-private partnership. However, certain emerging areas like cloud computing, social media regulations, cybercrime tracking has not been addressed in this policy. There has been no mention of safeguarding privacy of citizen by any law or policy. It is expected that major caveats of this policy will be rectified in the upcoming policy of 2020.

2.4 GLOBAL VALUE CHAIN GOVERNANCE

2.4.1 GLOBALIZATION

Before going to start what is global value chain governance and what is the importance of governance let me start with what is globalization. In simple term "globalization" is the interconnection and integration among world's economies, companies, culture, trade, and governments from worldwide [15]. The word "globalization" is not only connected with trade and business but also with political and social. So that one can define globalization in mainly three parts:

1. **Political Globalization:** It is the contribution or co-operation given by any government or any government leader worldwide. For example, "SAARC" summit is a group of eight different countries (Afghanistan, Bangladesh, Bhutan, India, Maldives, Nepal, Sri Lanka, and Pakistan). It is started from 8 December, 1985 and its main intention behind the formation is to establishing regional communication and association and studying the problems

like drug trafficking and terrorism. "ASEAN" group is another example where different country of Southeast Asia meets two times annually. The main purpose of ASEAN Summit is foreign ministers of different countries come together and talk about security, economic, political, and socio-cultural development of Southeast Asian Countries [16].

2. **Social Globalization:** It means sharing information, ideas, and cultural activities between different countries worldwide. For example, popular films and web series are best way to share ideas and experience. Social medias like Facebook, WhatsApp, telegram, etc., are best platform where one can share his ideas and experiences worldwide [16].

3. **Economic Globalization:** It means communication with different economy for trade and resource exchange. Economic globalization is best-known globalization today. For Example, McDonald's is the largest food chain with more than 37000 outlets in 117 countries. MNC's have too many branches at different locations [16].

2.4.2   HISTORY OF GLOBALIZATION

According to historians, economic, and social globalization can be existing to 320 BCE, with the establishment of Maurya Empire in India. At the time of Maurya Empire India was doing trade with Europe. Many historians have noted about "Silk Route" during second Centuries B.C. This Route was connecting China and Greece through Egypt, Persia, India, and Rome. China is trying to build "Silk Route" again with help of different countries and want to establish positions in different countries market. Many Religious organization or religious people from different countries like India, Europe, and Mesopotamia have established worship sites on foreign land. All imperial societies are involved in International trading by the tenth century B.C. [17].

International trade by the help of ocean was accomplished by the mid-1300s. Intentional and Unintentional trade of animal, crops, and plants help world to understand about different plants and crops. International exchange of agriculture product helps different countries to growth faster and sometimes it helps during drought. Due to International trade sometimes different types of dieses may happen due to communication between two different types of people coming out from different countries. For example, Black Plague. Black Plague is also known as "The Black Death." It was originated in central Asia and reached Crimea by the way of the Silk Route. It was mainly happened due to flies inside in black rats that traveled on merchant ships. 30% to 60% population of Europe was died due to "The Black Death" during 1347 to 1351 [18].

With colonialism Europeans were trying to develop complex regulations more quickly in the 1400s and it grew during 1600s and 1700s. Europeans captured and sold many slaves from Africa, America, and from many other countries. Slavery became main for cultural globalization [19].

In the middle of 1600s and 1800s, worldwide financial aspects depended on corporate greed and political impact was attached to the country's aggregation of tradable products and the size of the country's vendor's armada. A few nations were moving toward the act of protecting a nation's local ventures from remote rivalry by exhausting imports before the finish of the 1700s. During that time in many countries, independent movements began against imperial governments [19].

In late eighteen century, an economic and political system in America and Europe are controlled by private owners for profit. The colonial time frame in Europe and Asia declined step by step during the nineteenth and 20th century's, offering to up rise of new countries with exchanging connections to their previous colonizers [19].

World Bank formed in 1945 and its main purpose was to rebuild Asia and Europe after World War II. Its goal is to reduce poverty and improve living standards in people's life in low- and middle-income countries. In the end of 1945, IMF (International Monetary Fund) was founded and its main goal is to stable international financial system [19].

2.4.3   WHY DOES GOVERNANCE MATTER?

Value chain analysis with the help of governance and power clearly explains the distribution of profit and gains from trade. Governance of Global value chain is very important because on the basis of this process one can build "trade policy" for future and recent trades. For example, one cannot pay attention to R&D if only pays attention to production and marketing systems. This section clearly explains about why governance is important and what are the affecting factors to governance. The concept of "governance" is central to the global value chain approach. There are some factors recorded during the governance of global value chain and they are describing below:

1. **Market Access:** Sometimes developing countries cannot gain market access even if developed countries remove all trade barriers, because the chain in which producers feed into is many times governed by buyers. If any developing country's manufacturer wants to export product to any developed country then developer country's manufacturer needs to become lead firm in that chain. Decision of taken by lead firms sometimes lead to particular type of manufacturer and trader losing out. For example, recent studies show that small growers are marginalized in the UK–Africa horticulture (agriculture-related things, plants, etc.), chain. The reason behind this issue is small growers cannot implement strategies like big company (lead firm) and even small growers cannot have much access so that big company's strategies cannot fit better on these people and also large growers policy and strategies are influence by expectations of customer, non-profit organization and government agencies which deals with safety, environmental, and labor standards.

2. **Quick Acquisition of Production Capabilities:** Manufacturer who is act as lead firm in chain tends to find themselves difficult to learn. Leading firms are always very demanding in cost cutting, increase quality and raising speed. Sometimes local workforce is not capable to meet required conditions but lead firms teaches them and provide best practices and then local people can also contain skill and help in production. 1970s Brazilian shoe industry and 1990s Vietnamese garment industry are the best-known examples. A few ponders additionally show that brisk procurement of generation may make issue for structuring and advertising abilities.

3. **Distribution of Gains:** Examining the administration of global value chain one can see the dispersion of additions in the chain. Global value chain recommends that immaterial skills (R&D, structure, marking, showcasing) that portrayed by high hindrances of entry and command high returns can be rest by the ability to govern. Conversely, developing nation firms should focus on substantial action; pursue parameters set by the "governors."

4. **Leverage Points for Policy Initiatives:** GVCs are not only a relationship like string, but it may also be undermining government approach yet in addition offer new influence for government activities. Few chains are administered by lead firms from developed nations. These organizations give influence focusing on what occurs in provider firms of developing nations. This influence focuses are utilized by government and non-government organizations for raising work and condition benchmarks [20].

2.4.4   CHAIN GOVERNANCE

It is very easy to point out inter-firm relationships with the governance of global value chains. For example, the way by which UK supermarkets controls over their fresh vegetable supply chains. Chain governance not only specifies the type consumers want to buy (including different varieties, packaging, and processing) but also specifies qualities of product and type used for processing product. Firms can meet these requirements by auditing system and inspecting product. This clearly shows, governance in value chains has something to do with the exercise of control along the chain. Four key parameters affecting production process:

1. What manufacturer is going to be produced?
2. How product is to be produced?
3. When product is to be produced?
4. How much product is to be produced?

There is another parameter one might consider is price. Prices are variable and it is determined according to market. Governance structures become very important when transmit information about parameters. Governance shows the relationship between inter-firm relationships. Governance also shows the institutional mechanisms through which coordination is achieved in chain [20].

There are mainly two types of global value chain:

1. **Producer-Driven Global Value Chain:** Firms which control key product and process technologies set key parameters. For example, car manufacturing business.
2. **Buyer-Driven Global Value Chain:** Retailers and brand-name firms which focus on design and marketing set key parameters. For

example, agriculture, footwear toys and garments, etc., (Figures 2.5 and 2.6).

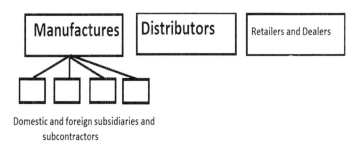

FIGURE 2.5   Producer driven global chain.

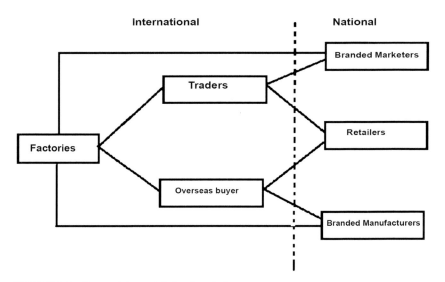

FIGURE 2.6   Buyer-driven global value chain.

From analyzing one can comprehend that in the worldwide economy there are two initial parameters: the producer decides what product will be created and how the item is to be produced is set by purchasers. These parameters values can vary. Buyers provide details and specification about products. Buyers can also set design of product for producer and process parameters too. There are mainly two cases in chain. One in which buyers give merely information about product and another in which buyer provide precisely information about quality and process standards. If buyer gives information about product to supplier then buyer becomes "lead firm" in the chain. In some cases, process, and product parameters are set by external agents. These external agents may be government agencies or maybe non-government agencies [20].

KEYWORDS

- **cybersecurity**
- **denial-of-service**
- **electronic information exchange**
- **electronic subsidizes move**
- **e-service**
- **global value chain governance policy**

REFERENCES

1. ITU, (2014). *Manual for Measuring ICT Access and Use by Households and Individuals.* Available at: http://www.itu.int/dms_pub/itu-d/opb/ind/D-IND-ITCMEAS-2014-PDF-E (accessed on 3 November 2020).
2. https://www.itu.int/en/ITUD/Statistics/Documents/publications/misr2017/MISR2017_Volume2.pdf (accessed on 3 November 2020).
3. (2014). *e-Government Survey.*
4. https://smallbusiness.chron.com/briefly-explain-evolution-ebusiness-62060.html (accessed on 3 November 2020).
5. https://www.statista.com/statistics/288487/forecast-of-global-b2c-e-commerce-growt/ (accessed on 3 November 2020).
6. https://www.webfx.com/blog/general/the-evolution-of-ecommerce/ (accessed on 3 November 2020).
7. https://www.researchgate.net/publication/263230383_The_Evolution_and_Development_of_E-Commerce_Market_and_E-Cash (accessed on 3 November 2020).
8. https://www.statista.com/statistics/273550/data-breaches-recorded-in-the-united-states-by-number-of-breaches-and-records-exposed/ (accessed on 3 November 2020).
9. *Cybercrime Magazine.* https://cybersecurityventures.com/cryptocurrency-market-watch/ (accessed on 3 November 2020).
10. *Accenture Cybercrime Study.* https://www.accenture.com/us-en/insights/security/cost-cybercrime-study (accessed on 3 November 2020).

11. https://www.whitehouse.gov/wp-content/uploads/2018/09/National-Cyber-Strategy. pdf (accessed on 3 November 2020).
12. https://www.merics.org/sites/default/files/2019-08/China_Monitor_20_Cyber_ Security-National_Security_EN.pdf (accessed on 3 November 2020).
13. Xi, J., (2014). *No National Security Without Cybersecurity:* ©*1Xinhuawang* 新华网. "习近平:把我国从网络大国建设成为网络强国" [Xi Jinping: China must evolve from a large internetnation to a powerful internetnation]. http://news.xinhuanet.com/ politics/2014-02/27/c_119538788.htm (accessed on 3 November 2020).
14. https://meity.gov.in/writereaddata/files/downloads/National_cyber_security_policy- 2013%281%29.pdf (accessed on 3 November 2020).
15. *Wikipedia, Globalization.* https://en.wikipedia.org/wiki/Globalization (accessed on 3 November 2020).
16. tutor2u. *Types of Globalization.* https://www.tutor2u.net/politics/reference/types-of- globalisation (accessed on 3 November 2020).
17. Silk Road. https://www.history.com/topics/ancient-middle-east/silk-road (accessed on 3 November 2020).
18. Black Death. https://en.wikipedia.org/wiki/Black_Death (accessed on 3 November 2020).
19. *History of Globalization.* http://connection.ebscohost.com/businessfinances/globaliza- tion/history-globalization (accessed on 3 November 2020).
20. John, H., & Hubert, S., (2001). *Governance in Global Value Chains.* IDS Bulletin 32.3.

CHAPTER 3

Role of Digital Education to Curb Gender Violence

DEEPANJALI MISHRA[1] and MANGAL SAIN[2]

[1]KIIT University, Bhubaneswar, Odisha, India,
E-mail: deepanjalimishra2008@gmail.com

[2]Dongseo University, South Korea

ABSTRACT

In one of the shocking and bizarre incidents recently, an eight-year girl, Asifa was gang-raped by a group of men before she was killed in Kathua, located in Jammu and Kashmir. This incident created huge protests on social media, and many celebrities, as well as common people, took to social media to display their anger and disgust. Violence against women takes place in various forms which cannot be confined to one state or a nation. It is not bound by any religion, caste, or social status. Without anyone's knowledge, she gets heartbroken and falls mentally sick sometimes leading to suicides or even murder thereby affecting her physical and mental health which ultimately leads to disastrous consequences on children and family. Rights of a woman are restricted and she is confined within the vicious circle of blames, humiliation, and segregation. Various studies, research, and surveys are being conducted on regular basis on gender violence. With the advancement of technology and development of the smartphones, women are falling prey to online violence at an alarming rate. The studies show women are victims of various types of violence and social media contributes to an increasing rate of violence against women falling prey to trolling, body shaming, etc. This chapter would aim at making a theoretical study on how women fall prey to social networking sites and remedial actions to be taking for preventing such measures.

3.1   INTRODUCTION

After the Nirbhaya rape case which created havoc all over the country, the rape case has been on a rampage. Recently, a girl named Twinkle Sharma, minor who was just three years was brutally raped with her body parts mutilated which was found from a garbage bin caught attention of general public. Such cases have become quite common and not a single day passes by without such cases being presented by the media. In another yet shocking and bizarre incident recently an eight-year girl, Asifa was gang-raped by a group of men before she was killed in Kathua, located in Jammu and Kashmir. This incidence created huge protest in the social media and many celebrities as well as common people took to social media to display their anger and disgust. In the age of social media almost all incidences catch media attention and can be easily accessed by people irrespective of their age, location, and education level. This it is very good that awareness is made easily through social media. However, sometimes-social media can lead to drastic situations like in 2007; a teenage girl Megan Meier committed suicide because she admired a guy on MySpace who was actually her classmate's mother using pseudo name. When she suddenly ended the friendship, Megan went into depression and took this drastic step of suicide that same day itself. The results are widespread. According to USA's Mail Online's report by Jack Doyle as many as 12,300 cases are linked to the site. A crime is reported to police that's linked to Facebook every forty minutes. Facebook was referred to in investigations of murder, rape, child sex offenses, assault, kidnaps, and other crimes. The Mail Online also reports about the most talked Ashleigh Hall murder case that happened primarily because of Facebook. The case goes as such, according to the website Ashleigh Hall who was a teenager fell in love with one Peter Chapman via Facebook Ashleigh Hall was murdered by serial rapist Peter Chapman after he befriended her on Facebook. Chapman, 35, displayed himself as a handsome teenager with the name Peter Cartwright to draw in 17-year-old trainee nurse Ashleigh and trapping her. He texted her continuously and planned to meet her some weeks later, claiming to be the father of Peter in-order to explain why he looked nothing like his photo. Chapman drove Ashleigh to a lonely place in Thorpe Larches, which was nearer to Sedgefield in County Durham. Once there, he forced her to have sex with him and then he bonded and gagged her with duct tape, wrapping so much around her head that she was suffocated which resulted in her death. He then threw her body in a ditch and disappeared. Chapman is now serving sentenced 35 years in jail after being sentenced in 2010 for Ashleigh's kidnap, rape, and murder." Singer Sona Mohapatra was very embarrassed when she was victim of trolling for her post on the famous Bollywood star Salman Khan where she had written about the actor's crime related to killing people and shooting a blackbuck. She wrote, "*Women thrashed, people run over, wild life massacred and yet #hero of the nation. 'Unfair.' India full of such supporters*"—Sona tweeted [1].

In September 2014, a Pennsylvania teenager was jailed for two years after he posted pictures of himself having oral sex with a statue of Jesus on social media. The images went viral and faced a backlash among communities, New York Daily News had reported at that time. Earlier this year, the video of a Columbus girl who was per scoping her boyfriend rape another girl was all over social media pages, and then the video Baton Rouge police who shot and killed Alton Sterling. All of these are just some examples of what's playing out every day for kids to watch.

Advancement in technology has led to the massive use of social media added to the creation of smartphones and its application has allowed people to access social media virtually anywhere. They have increased our ability to communicate invariably with large number of human beings at any time of the day or night. Technological advancement and rampant use of social media has made dissemination of information easy, fast, and efficient. Simultaneously, it allows the sender as well as the receiver reciprocate or show their reaction through silence or may opt for sending a message in anonymous mode too. Facebook, Twitter, Instagram, LinkedIn, YouTube, etc., are some of those social networking which are available to hundreds of millions of people all over the world and have successfully united people across the geographical boundaries that are like-minded, share a common topic, express feelings which is reciprocated by supporting them. In any event related to a community, or sometimes we see a post which has the features of a missing child. We love to update our profile picture post something related to our feelings, emotions we post any video which attracts us, or some blog, and see how many reactions we have received, and it reaches our contacts, friends, friends of friends and even unknown people with just a click of our mouse. Social media sites have given people ample options to create awareness regarding an issue, making our voice reach to a vast audience, and viewers, we can also invite others to join if they are empathetic towards our cause and share same feeling as we do.

Internet plays major role in professional as well as personal lives of a huge number of people. In addition, it has dramatically taken a significant surge since the last decade and of late people have started using internet for shopping, looking for schools worldwide to get their children educated, looking for options online for entertainment, work, and jobs. Some even work online and get paid online!! Social Networking Sites are taken to be a social and professional platform to flaunt latest pictures, albums of events, experiences among friends and others.

Today's media has undergone remarkable transformation by connecting people involving them and engaging them in a common platform as per their tastes and interests. Due to the liberty provided by the social media, Women consider it as opportunity to feel free in bringing out certain issues that they have either faced in their lives or they have seen it happen like domestic violence, sexual violence, etc.

Social Networking Sites have not only created a forum of texting and sharing and knowing people around the globe, as it was considered to be a platform for displaying personal achievements but also it has been developed as a technique for creating a platform for political agenda just before few years back. Like For example, in the words of Esfandiari [2], the political unrest that took place in Iran in the year 2009 elections, it was more or less aided and aggravated due to Facebook. This was the reason why Iranian Government made an announcement terming Social Networking Sites as 'hidden enemies' which ultimately led to its ban countrywide [2].

Social Networking Sites have grown and developed rampantly over the years and presently these Social Networking Sites as well as sites have an eye on some countries, people who are members of those groups which are targeted and some professionals belonging to a specific profession. Though there are also instances of some Social Networking Sites which are very successful and mushrooming their business, but they have a very small number of members, which may range from some hundreds. The size of members in social networking sites depend the frequency of group member's interaction, finance, and even the strategies these sites employ to retain their members. Social networking sites can be defined as those web sites which permit their users to create a public profile or even a customized profile that professionally coherent the relationship with other members in such a way that the profile can be accessible by any other user who search their file.

The Indian scenario is no different. A majority of adolescents in the age group 12–19 in India regularly uses digital technologies like smart mobile phones with internet. Various service providers like Jio, Airtel have slashed the data rates which makes the users to have at least one phone with them. By just pressing a fingertip they go to the virtual world. No doubt they use internet to work for their project or search and access study materials, it also facilitates them to create an identity of their own in the virtual media They open account on these Social Networking sites without being fully aware about its impact. They try to develop friendship and build relationship with whom they know and also sometimes with whom they do not know. It was predicted that internet would internet may be used as an aid to beat the un economic, social equality but now the situation has become such that the adolescents are fast becoming victims due to the internet use. Sometimes the service providers of social networking services lure the users with a promise of lucrative rewards by providing their real identities which includes gender correctly. These users fall prey easily to this trap.

3.2 LITERATURE REVIEW

For instance, the concept of cyber-bullying which was unknown has become a regular phenomenon of late. Cyber bullying means when messages are sent by unknown people through e-mails or posts, videos, leading to embarrassment in order to threaten the victim. Van Laer (2013) talks about the hazardous impact of cyber-attack as "emotional distress. Withdrawal from social network sites or even life itself." [3] At the same time, Görzig and Frumkin [4] comment that "cyber-bullying may appear in different forms as seen in sending unwanted, derogatory, or threatening comments, spreading rumors, sending pictures or videos that are offensive or embarrassing by text, e-mail, chat, or posting on websites including social networking sites" [4].

3.2.1 *GENDER BASED VIOLENCE AND SOCIAL NETWORKING SITES*

Bloom [5] defines gender-based violence (GBV) "as a very common concept which is used to explain the violence that takes place resulting from the normal role performance expected of a person that is related to a specific gender which also includes inequality in relationship of power between both the genders with reference to the social context of a family" [5]. Though it is considered that women care victims of gender violence, but in today's changing world, there are many cases where men, boys have been victimized. Bloom again talks of violence of violence against men. He adds that one cannot deny that violence against men does not happen. He exemplifies that men easily become target related to physical as well as verbal attacks due to the pre-existing conception related to their gender. Young boys are often tortured sexually and physically abused by the same sex adults. Thus men and boys are victims more often within the confinement of their family sometimes by their partners and sometimes by their own children.

However, the focus point is the gender violence that takes place with women and violence is inflicted on in all forms with the girls and women of all age groups because all the Human rights reports and other databases clearly mention with the acknowledgment that women and girls are worst victims of gender-based violence. To make matter worse, women, and girls suffer due to the gender discrimination right from their birth because a boy child is preferred in a family. The United Nations Fund for Population Activities or UNEPA sums up this scenario as Women and girls especially who adolescents and women are the primary targets. Not only that, the risk of women and girls being more prone to gender-based violence, they are more likely to suffer intensified result comparable to their men. Due to gender violence and the lower economic class women are left with less choice as well as few resources with them in order to come out of this miserable plight and fight for justice due to which they suffer miserably which has worse impact on their reproductive health like unwanted pregnancies and forced to go for abortion which are unsafe [6]. This causes drastic effect on their health and is at a higher risk of getting sexually transmitted diseases and HIV which results in being discarded by their family members. It is a well-known concept that women and girls face violence during every stage of their life cycle. In some cases, women face a series of sequential cases related to violence which may start right from the moment when they were born, which may continue during their adolescence and till the time they are adult though there have been some cases where women are abused in their old age. Most adults and adolescents use social media and use it regularly. The concentration of users on Facebook makes it easy to spread information on that platform.

The statistical data shows that there have been maximum cases where women experience violence during the adulthood and lowest when they are children. This data has helped to understand the impact of violence that is inflicted on them in different stages of their life. No doubt it has drastic consequences on their issues related to their physical health as well as mental traumas. Twitter conducted a study on the Gender Based Violence in the year 2014 in order to make an extensive analysis of the use of misogynistic language in developed countries. It was found that the UK and US has maximum number of users who use misogynistic language. Though some conclusion yield same outcome as another survey which focused only on the United Kingdom, the specification methods used by them is not satisfactory and are full of shortcomings with regard to location, identification of gender and the language features. In the year 2014, a survey was conducted to analyze the relationship between a misogynistic language and the statistics of rape in the US. This study was carried out through collection of a broader data like the harmful effects of sexual violence, etc., not just the type of misogynist language used by the users. They established a close connection between the violence and the growth of cell phones. It was found out that the growth in mobile phone use has provided the opportunity for increased access to the internet as a result of which one can easily get access to social media. It is not unknown that social media is a platform of sharing and exchange of useful information on one hand and at the same time it gives rise to drastic situation. As per the study conducted through search in internet through Google Search engine as well as and Google Scholar along with the information that is accessible online to the users who are writers, media persons too were involved in the study. Out of which five cases were selected that were reported and was found that in those five cases, the victims had primary contact with their attackers on Facebook which means they knew each other well. In all the cases the attackers were men who were senior to the victims who were in the age group 17 to 20 compared to 24–34 years. The victims witnessed violence of all kinds of physical abuse, mental traumas, isolation, and also economic violence. In one case, the victim gave up and she died, whereas in the other two cases victims were hospitalized because of acute psychological torture. In the first three cases, they were constantly raped where as in the other two cases, they were molested, and the attackers tried to rape them. Gender-based violence have resulted in many cases where the victim as well as the attackers are known to each other and the education of the victim is the only necessity in order to improve their situation.

The crimes that are committed against women constitute one out of five which make 19% of all crimes that are committed. The scenario is horrifying because it is more than the other types of crimes and it is more than terrorism and fraud. According to the police reports this year about 1,347000 women were victimized due to some reason or the other and it was social media that has actually aggravated the gender violence. Off late body shaming has emerged to be a new form of sexual violence where a practice is followed by the netizens to humiliate any user through mocking or making critical comments related to their body structure. Nowadays while a user scrolls through news feed or viewing an Instagram, it is very usual to see a nude picture. In many cases, it is done in order to take revenge from an ex girlfriend or if the girl denies to be in a relationship with someone. It would really be shocking to see a nude picture of a 21-year-old girl and a 60 years man in intimate position. The attacker who posted the picture alleged that he did it because he wanted the entire world to know that his girlfriend was cheating on him. Technological use in harassing someone makes the victim more traumatized. In another case, a girl committed suicide and she posted it on a networking site live because she was a victim of body shaming as user teased her as plump. Therefore, in all these cases all the victims are women and online crimes are rapidly growing at an alarming rate.

The survey that was conducted on social media showed that a vast number of respondents have knowledge about sexual violence and also know the association between social media and sexual violence. It was found that all the respondents were acquainted with Facebook and did not have much idea about the other social networking sites and nearly half of them said it is Facebook is a hub of abuse and harassment. Though some of them said they knew Twitter, Turnbir, Instagram, and opined that sexual violence can take place in the form of texting abusive messages or harassing online to the victims.

It depicts the breakdown of awareness of sexual violence related to social media: 75% of respondents knew that social media can be used as a tool which would control, trolling, stalking, with the existing or past boyfriend 74% were of the view that they knew that users share others posts or even personal pictures without their consent 63% had knowledge that users post or share sexually violent texts or pictures 52% had knowledge that social networking sites are used for exploiting and finally it was 10% respondents who claimed that they did not know that harassment, trolling, and humiliation take place on Social Networking Sites. A survey was carried out with a group of youngsters in the USA consisting of a group of educators, lawyers dealing with cyber violence and workers between June and July 2013 and the objectives of the study was:

a. Collecting data related to gender violence and social networking sites;
b. Identifying the prevailing strategies of prevention;
c. Finding the steps for improvisation for prevention programming.

The survey was conducted online where a link and letter of invitation was sent via e-mail to a group of respondents in the USA, comprising of various organizations, various educators, lawyers dealing with cybercrime, internet service providers. Individuals were encouraged to forward the information to community networks with the aim of increasing the survey's reach and to allow individuals to identify relevant local partners and 25 contacts. 10 questions were given to them. It should be noted that many respondents fit into multiple categories (for example, a community health center may also provide sexual assault support services), so this represents an approximate breakdown only. In the survey, approximately 25% of respondents directly noted that they work with children, adolescents/teens, and/or youth. Some respondents who were contacted did not perform the roles as a youth. They served major roles in coordination with their parents, were students studying in universities, and others who claimed to have seen or know the cases related to cyberbullying taking place.

3.3 HOW DO WOMEN FALL PREY ON SOCIAL MEDIA?

Social Media has become more vulnerable and accessible due to smartphones, laptops, and low-cost service providers of internet, messages could now travel so fast and multidimensional. People have taken it to be one of the most reliable methods of communication. The users know very well that one can maintain anonymity and the message can reach a larger number of audiences. Therefore, users try to post rumors, and criticize someone whom they know or not because they want a very vast audience without just confining to their contacts thereby creating a new level of torture to the victim. Liou [7] states that "social media campaigns are less effective when conducted as standalone activities, compared to when integrated with face to face and on the ground activities" [7].

Kabir Shirgaonkar, CID of Goa police, tried to explain the students about certain concepts related to cybercrime. Online crimes like Hacking, Spoofing, etc., are quite prevalent in the virtual world and therefore he appealed the students to be very cautious because of its rapid growth. He expressed his fears that very soon cybercrime will be the most commonly committed crimes compared to all conventional crimes because the growth of net users are increasing day by day. Shirgaonkar was invited to deliver a talk on Cybercrime organized by traffic cell of the school students during a summer camp organized by traffic cell, Calangute, at St Joseph's High School, He cited many examples where women and school girls fall prey to the Social Networking sites, He emphasized and strongly advised the girls to be very cautious while using social networking sites and dissuaded them to be friends with someone whom they did not know Facebook. He also told them not to give their true details to unknown people because there were so many users waiting for this opportunity and once they get the pictures of women and girls, they morph them and put them on various social networking sites. Then they start blackmailing the girls and thereby destroying the reputation and humiliating them on internet. He also said that online crimes are recorded to be about 30 to 40 cases of *GBV*.

There have been enormous reports of digital intimate violence that is taking place in various Social Networking Sites. Facebook, one of the most popular networking sites has portrayed a feminist image for itself. Its executive and author, Sheryl Sandberg, is highly appreciated and is considered to be the ace feminist, Betty Friedan because she urges women to bring a complete metamorphosis to their lifestyle. They need to work hard and struggle in order to improve their sorrowful plight. Social media has been often branded to be most perilous place for becoming a dangerous place for those women who are financially weak, and downtrodden. No doubt Sandberg talks of economic liberation, improved lifestyles sounds like the language of the elites. Her motivational talks explore a new area for development. Her narration is highly influencing which ranges from being a student of Harvard, to her career in Google and now as executive at Facebook. She is a skilful entrepreneur and is tech savvy with many elite and educated co-workers to guide her and work with her. She has a very supportive husband, good male colleagues and definitely cannot be considered as one of the subjugated women.

There are many forms of violence against women that usually take place through social networking sites which are discussed in subsections.

3.3.1 TROLLING

Trolling causes frustration and mental tension for the users of social media users. When one posts something on Twitter, people comment on her dress, attitude in a very negative way en-masse. This is known as trolling. Online trolling as a constant uninterrupted abnormal behavior performed by any user or a group of users [8]. Trolling can be considered as very offensive and it is an attack where the attacker is an online user of social media causing problem to the victim by posting certain picture, post, comments intended at the victim. Trolling is of different types like those that insult trolls, unrelenting debate trolls, vulgarity trolls, all-caps trolls, flaunting trolls, etc. Extensive research has been conducted for business undergraduates regarding any kind of case related to psychological nature and crimes considering the fact that they were the future leaders using the social media. It was found out that the students access internet very actively and they use many service providers of internet. It was also found out that the time spent on social media led to their trolling and also if they are women then frequency of trolling is more.

3.3.2 CYBERBULLYING

Cyberbullying is another form of internet violence against women which has taken its course because of development in technological revolution making communication much faster and accessible. It is a form of threatening someone or harassing her by any individual or a group of online users. *Youth Internet Study Survey* conducted a couple of studies which verified the information that was taken and it was revealed that the youths who have been bullied online constitute 9%, and the male youths who confided that they had actually made indecent remarks comprise of 28% to victim which actually has increased from 14% last year and 9% of youths confessed they used online to harass a victim with whom they were obsessed to. This was actually an increase of 8%. This survey gives us a clear picture that Social Networking Sites are used to bully and harass someone [9].

This leads to an argument that Social media is responsible for online harassment and these sites should take the blame. Some intellectuals opine that the young users should be banned from using internet which would put a limit to cyber crimes. On the other hand some others viewed that social media cannot be blamed for this because it depends on individual who is using social networking sites as means of harassment. Therefore, one solution could be using social networking sites to fight online harassment and stop crime against women and educating the users of internet. These are certain instances how cyber bullying is done by the attackers on social networking sites to humiliate and put mental pressure on others. These are some of the excerpts which are posted by from Help Phone online forum for children and teens which clearly show how the teenage girls and kids are bullied.

There was a user who uploaded a text on her Facebook account with disgust that it would be a bad year for her because there were many students in the same classes. One user wrote back why she hated him even though he was nice to her. There were so many examples on Facebook where one is trolled for some reason or the other. If one posts her opinion about an actor, her academic degree is questioned; if one post anything about a politician, her profession is questioned. All these examples clearly show how the social networking site is used by the attackers to torture the victims through derogatory comments and giving threats to them. It should be emphasized that committing violence in the form of threats, is a criminal act and could convicted in the court.

3.4   METHODS TO CURB GENDER VIOLENCE THROUGH DIGITAL EDUCATION

3.4.1   *EDUCATING WOMEN FOR THEIR SAFETY*

The girls and women should be made aware and study the rate of crimes taking place due to gender violence. The women and girls of all age group should be oriented by the social media experts regarding its usage, privacy, and safety. The girls should know how to block and delete accounts and learn to distinguish between fake and real accounts. Training should be imparted to the girls regarding the preventive methods using social media. Awareness should be created about the usefulness of Digital Education to curb crime.

As we all are aware of the proverb, prevention is better than cure, same thing is applicable here. Efforts to stop violence needs to be taken before it starts. One of them is primary prevention which comprises of a holistic approach ending violence and to promote non-violence and peace. Response may refer to different kind of measures which are taken to support and protect the women who have suffered and experienced violence. Primary prevention aims to identify and address the underlying reasons of violence to minimize the opportunity whenever it happens in the first place which means it is an effort to prevent violence before it crops up. The attempts proved that for a campaign to be effective, several considerations are to be thought of in terms of planning and implementation of the awareness. In terms of good practices, the three campaigns found that for a campaign to be effective, utmost care needs to be taken to plan and to implement the campaign.

We all are aware that social media is a means of connecting people at every level through professional as well as non-professional interaction and getting people communicate with each other at a larger scale. Social media tools or services utilize the internet to assist conversations. It may also include web-based and mobile technologies which are used to convert a casual communication into interactive dialogue, which is also a platform where people can voice their opinion in various ways. Social media empowers people to engage with other people.

In order to prevent gender violence, lots of social media campaigns have been launched that includes all the elements to generate chances of getting best results. At the same time many successful social media campaigns are organized and events are conducted which can be entertaining. Other good practices that also emerged from the three campaigns which are to intended at creating an on-the-ground community which feels a belongingness of being the owner of the campaign and secondly, working with this community to assemble more members online as well as offline can be found to be one key to ensure success. The crusaders try to develop online activities that are not only entertaining but also easy to participate. It helps to make personal connection more appealing to people's emotions. One pragmatic approach could be to hold online discussions and debates around current attitudes and behaviors on gender behavior, bringing equality between gender, giving respect and maintaining healthy relationships, and advocating for a model with positive gender-equitable behaviors. Incentives can be given for successfully engaging audiences ranging from capacity development opportunities and for recognizing material rewards. Lessons learned from the campaigns highlight the limitations of social media for the prevention of VAW. This discusses the fact that social media campaigns are the least effective when it's conducted as the activities with a single individual as compared to when they are conducted with a group in the form of face-to-face discussion or on-the-ground activities. It can be like to understand and then measure the type of impact social media interactions have on the general audience which may be difficult, yet it's possible with the right monitoring plan. A general overview suggests that social media are more prone to be accessed by a huge audience to the messages. It can also be available to a much smaller number of people in learning activities, and an even a less number of audience can take tangible actions to prevent VAW.

These campaigns describes lessons learned from the three campaigns in terms of what changes for VAW prevention a social media campaign can contribute to. The campaigns found that social media is useful in terms of mobilizing people to provide a space where groups of people can take initiative to influence changes aimed at preventing VAW. The campaigns also found that social media can be used to strengthen networks, promote relationships of being part of a community, and help create a con existential and healthy environment. Social media can provide a space for dialogue that would not otherwise be available.

All the effective social media campaigns use the social media platforms to combine them with reputation, reward, and influence based contests or challenges. The offline components to gather and engage the youths around the campaign by rewarding both the kind of people who contribute as well as who benefit. They partner on the ground in the aforesaid target zones and communicate clearly about the campaign's substantial results to all the parties who are involved. They don't merely propagate information and are clear about the results. The people who want to create a difference, be rewarded, and recognized for it are the ones who require building in virility. There are other methods which can help curbing gender violence.

All identifying information which identifies self or others should be strictly prohibited and if it's published, then there should be with clear-cut definition whether it should be visible public or to private audience. There should be minimum intermediary's obstacles to download pages, posts, or content pertaining to privacy concerns and people would refrain from it if there is a warning message followed by threats. Companies should take action and there should be clear accountability measures which necessitate clarity in response to complainant. A case surfaced in Pakistan where a Pakistani blogger, under the name Baaghi, was highlighted because his national identity card, marriage certificates, even residence of last 10 years including other such private information was updated online which finally ended in an attempt to assassin the blogger.

Similarly a woman in Sarajevo which is a part of Bosnia and Herze-govina tried to report about a fake Facebook profile that was created in order to damage her reputation. For that she tried to take help from an organization called 'One World-See' to report that the profile and the forms were only available in English. That organization assisted people like her in promoting mutual legal assistance treaty (MLAT) which reforms to increase assessment to get justice in all the cases of technology-related VAW which provides clarity along with transparency and accountability regarding any action on the content in the internet and privacy requests. Many women reported that they get very few response or even if they get, it's a just an automated response. By this way it is more transparent and is more accessible as well as accountable to public in the workplace inside the departments and all the staffs who are responsible for responding to the internet content and privacy complaints that they make.

3.5 CONCLUSION

No doubt report links are developed, no doubt common platforms are formed to combat the violence against women due to social networking sites, yet it would continue posing challenges to the victims and explore new opportunities for the officials of cyber crime as it is conventionally perceived that social media is dangerous for women giving rise to an increased number of crime. Even though restrictions are made, security system is improved, yet the attacker may find new ways of harassment. With the consolidated presence of social media in our lives, one needs to explore the unexplored strategies' in order to understand it and accessing it. Instead of branding social media as 'unsafe,' one need to understand it getting its benefits and at the same time, it should stay away from its negative impact that is related to cybercrimes [10]. Without ignoring the fact of the existing gender violence due to social networking sites at every level of age group and in every sphere of the society, hegemonic masculinity as well as neoliberal ethics ought to be questioned now flow seamlessly across both public and private spheres. Therefore, while addressing the topic of violence against women, various aspects related to culture, class, gender, unawareness, etc., should not be ignored where the virtual platform shames, humiliates, and blames the victims due to the innocent choices they make in their personal lives.

KEYWORDS

- **digital education**
- **gender-based violence**
- **hegemonic masculinity**
- **mutual legal assistance treaty**
- **social media**
- **social networking sites**

REFERENCES

1. Sona, M. *Women Thrashed, People Run Over, Wild Life Massacred & Yet #hero of the Nation.* Unfair. https://twitter.com/sonamohapatra/status (accessed on 3 November 2020).
2. Esfandiari, G., (2010). *Iran Says Facebook and Twitter Are Country's 'Hidden Enemies'.* Available at: http://www.rferl.org/content/Iran_Says_Facebook_And_Twitter_Are_Countrys_Hidden_Enemies/2171343.html (accessed on 3 November 2020).
3. Van, L. T., (2014). The means to justify the end: Combating cyber harassment in social media. *Journal of Business Ethics, 123*(1), 85–98.
4. Görzig, A., & Frumkin, L., (2013). Cyberbullying experiences on the go: When social media can become distressing. *Cyberpsychology: Journal of Psychosocial Research on Cyberspace, 7*(1). doi: 10.5817/CP2013-1-4.
5. Bloom, S., (2008). *Violence Against Women and Girls: A Compendium of Monitoring and Evaluation Indicators.*
6. Rodriguez, E. M., (2016). Gender violence and social networks in adolescents: The case of the province of Malaga. *Procedia-Social and Behavioral Sciences, 237*(2017), 44–49.
7. Liou, C., (2013). Using social media for the prevention of violence against women: Lessons learned from social media communication campaigns to prevent violence against women in India, China, and Vietnam. *Partners for Prevention.* Retrieved from: http://www.partners4prevention.org/resource/using-social-media-

prevention-violenceagainst-women-lessons-learned-social-media (accessed on 3 November 2020).

8. Axelrod, E., Tsemberis, E., & Siegel, S., (2012). *It's Time for a Woman Moderator: Equality in the 2012 Presidential Debates*. Retrieved from: http://www.change.org/petitions/it-stime-for-a-woman-moderator-equality-in-the-2012-presidential-debates (accessed on 3 November 2020).

9. Florea, M., (2013). Media violence and the cathartic effect. *Social and Behavioral Sciences, 92*, 349–353. doi: 10.1016/j.sbspro.2013.08.683.

10. Fellow, A. R., (2010). *American Media History*. Cengage Learning Hutchinson, T, 2008, Web Marketing for the Business, Blackwell Publishing.

11. Young, J., (2012). *The Current Status of Social Media Use among Nonprofit Human Service Organizations: An Exploratory Study.* Richmond, VA: Virginia Commonwealth University. Retrieved from: http://hdl.handle.net/10156/3775 (accessed on 3 November 2020).

CHAPTER 4

Cybersecurity Threats and Technology Adoption in the Indian Banking Sector: A Study of Retail Banking Customers of Bhubaneswar

SUKANTA CHANDRA SWAIN

KIIT Deemed to be University, Bhubaneswar, Odisha, India,
E-mail: sukanta_swain@yahoo.com

ABSTRACT

With the insistence of digital mode of transaction, while the banks are having conveniences in the form of customer management and revenue accretion through commission and convenience charges, a group of customers have been practicing it with mixed reaction, and the rest are considering it as the most bitter-taste medicine that makes them away from the technology. Moreover, owing to changing norms such as enhanced minimum balance in savings account, restrictions in withdrawal frequency from ATMs, misappropriating the amounts in any defunct savings account and apprehensions from FDRI Bill-2017, the customers' confidence on banks has been declined tremendously. As such banking industry is struggling with multiple incidents of banking fraud, cyber threats and NPA issues. Adding to this, low level of customers' confidence has genuinely put question on adoption of technology and sustainability of the Indian banking industry. On this backdrop, this chapter aims at unfolding how technology adoption and sustainability have been a challenge now before the banking industry through direct personal interview of 100 retail-banking customers based at Bhubaneswar. Descriptive statistics is used to analyze and interpret the collected data.

4.1 INTRODUCTION

Competition and change are two such features of any business in today's world that are making a business perish if found unsuitable to cope and take advantages amidst odds. Banking industry in India is no exception to this trend. In recent years, the industry has gone through drastic changes in banking norms and since the advent of the private players in the industry, there has been intense competition. Banking industry in India, unlike other industries, perform with limited autonomy imposed by the monetary authority of the country. Indian banking sector has passed through a drastic change in the modes of its operation-from traditional mode of banking to convenient mode of banking. Of late, Government of India is also insisting digital modes of transaction and thus endorsing digital banking. Emergence of digital banking has close bearing with the initial reforms of 1980s and 1990s in banking sector in the form of Computerization of banks and Core Banking Solution that had enhanced the convenience of customers by providing them anywhere and anytime banking. Although, as usual, the bank employees were having some sort of reservation for inconvenience caused in adoption of technology in banking operations, they got acclimatized and adopted the new technology slowly and steadily. Adaptation of digital mode of transaction has made the banks advantageous in the context of customer management and revenue augmentation. Digital mode has made the job of the banks easy in managing customers as some of the customers do most of the transactions without visiting the bank branches. Similarly, by collecting commissions and convenience charges for some of such transactions, banks have also succeeded augmenting their revenue. However, from customers' perspective, digital mode of transactions are yet to be accepted largely owing to word-of-mouth publicity of some erratic experiences of a few customers and lack of knowledge and confidence of those who have not yet adopted the digital mode of transaction. If most of the population doesn't accept the digital mode of transaction, then the dream of making India a Digital India will be a daydreamer only. Those who have been using the digital mode have undoubtedly the experience of convenience but at the same time, they also have some hard fact to unfold that may reveal the dark side of the noble jingle-'Digital India.' While thousands of positive words of mouth are required to establish the positivity of a product or service or new technology, a single negative word of mouth may be enough to make it a failure. Cyber threat is one of those negative words of mouth for digital transaction that stands strong on the path of accepting the mode. Moreover, Statistics on cyber-crimes and India's readiness for cybersecurity strengthens the negative word of mouth pertaining to cyber threats and makes the motto of 'Digital India' seemingly unreachable. On this backdrop, this study has been taken up to find out whether the retail banking customers are aware of cyber threats and have experiences of any such incident. Moreover, this chapter also advocates policy prescriptions for security on the basis of respondents' (customers) feedback. For the purpose, 100 retail customers based at Bhubaneswar have been studied through direct personal interview and descriptive statistics is used to analyze and interpret the collected data. The finding of this study will insist the government to focus on cybersecurity and help the banks to strategize accordingly to win customers' confidence.

For having the easy and comfortable access of facilities for instant credit and debit, different digital modes for financial transactions are in offer. Those are plastic money, net banking and virtual wallet. Those modes seem to be convenient, easy to use, and secured. Credit Cards, Debit Cards, and Pre-Paid Cards are variants of Plastic Money that caters to easy and instant access to credit and money. An application on Mobile that offers automatic management of accounts or taxes and convenient electronic records facility for the users is called Virtual Wallet. Being installed in the Mobile, Virtual Wallet is immensely handy and accessible 24×7.

The intensity of usage and advantages to the users of the different modes of digital transaction are guided by the perception and inherent technology in it. Recent development in computing and mobile devices has changed the payment industry. While the supply side of the digital modes of payment is keen to experiment any advancement in technology-embedded modes, the demand side is bit skeptical about the same owing to negativity in perception and adverse word-of-mouth. New patterns have been evolved by financial organizations for augmenting the usage of different modes of digital transaction imbibed with advanced technology. Although enough research and development (R&D) are in place to identify potential clients, both online and offline, concerns of organizations pertaining to incidence and depth of usage of digital modes of payments are still there which put question mark on the profitability and sustainability of banking organizations.

Since existing client base is not enormous and there is stiff competition in banking industry, enrolling the masses in the financial system through financial inclusion is the only way out for sustainability of banking industry. Trial of new financial organizations, in this regard, in recent years, using Aadhar Card-biometric identity card is definitely gives a ray of hope for a bright future but these organizations are yet to go a long way in this direction to achieve the quibbling mass. All these attempts in recent years have escalated the demand for cyberspace exponentially. However, India is not so strong in cybersecurity for which people fail to appreciate and accept the digital mode slogan. Every now and then people are coming across news related to financial fraud pertaining to digital mode of payments. On this backdrop, this chapter is an assessment of the retail banking customers' perception and preference on transactions through digital modes such as Plastic Money, Net Banking, and Virtual Wallet Services vis-a-vis bank branches. In order to address the concern of customers on security threats while using digital modes of transactions, appropriate suggestions are made to beef up the security aspect entangled in all those gadgets. The scope of the study is confined only to Credit Card, Debit Card, net banking, and Virtual Wallet Services and would consider users of such contemporary banking appliances in Bhubaneswar city of India.

The plan of the chapter is as follows: Section 4.2 of the chapter precisely reveals the objectives and methodology of this chapter. Existing literatures have been studied and the gist of most relevant ones has been presented in Section 4.3. Technology in Indian banking sector and acceptability of the same by the stakeholders has been explained through technology acceptance model (TAM) in Section 4.4. Section 4.5 unfolds the danger of cyber threats in India factually. Data analysis and findings of the study have been presented in Section 4.6 and Section 4.7, respectively. Concluding remarks have been placed in Section 4.8.

4.2 OBJECTIVE AND METHODOLOGY

The objectives of this study are as follows.

1. To present the issue of cyber threats in Indian Banking sector that question the adoption of technology and sustainability of dynamism in banking industry.

2. To assess the acceptance of digital modes of transaction with the help of TAM that determines the sustainability of technology-oriented changes in banking industry.
3. To unfold the experiences and/or perceptions of the retail banking customers of Bhubaneswar (India) pertaining to cyber threats/ crimes in the context of sustainability.
4. To recommend what needs to be done at policy level on the basis of respondents' feedback so as to make the usability of the changes in banking industry sustainable.

Methodology adopted for achieving the above-mentioned objectives are as follows:

- **For Objective 1:** Secondary data have been collected from concerned departmental published reports and reports of the relevant Ministries of the Government of India.
- **For Objective 2:** While existing Technology Adoption Model (TAM) has been referred, inference regarding adoption has been drawn on the basis of primary data collected from 100-retail banking customers based at Bhubaneswar (India).
- **For Objective 3:** Primary data has been collected from 100 retail banking customers based at Bhubaneswar (India) pertaining to; a) whether they use any or many of different digital modes of payments, b) if they use any or many, what have been there experience and c) if they don't use any or many, what are the possible reasons for non-acceptance. 100 sample units have been studied through convenient sampling method.
- **For Objective 4:** The same set of data used for Objective 4 has been used to recommend the policy prescriptions.

4.3 LITERATURE REVIEW

Having based on reviewing existing literature, the present study is undertaken. The gist of most relevant literature is presented below.

There is massive potential for mobile payments (m-Payments) in India as there are a good number of mobile subscribers are there and a reasonable percentage of population studied has little or no accessibility to banking facilities. It is inferred that m-payments are the outcome of accessibility of low-priced handsets, minimal mobile call rates and massive data network coverage throughout the country [1]. There is strong inter-connectivity among readiness for adoption, perception regarding risk and intent for use of payments through mobile in India [2]. In order to make NFC-based application highly secured, there must be checks in attacks like data insertion, relay attack, eavesdropping, data modification, and denial of service (DOS) [3].

As a new short-range wireless communication technology, NFC enables uncomplicated, secured, and spontaneous peer-to-peer connection between NFC-enabled devices. Under short-range radio technology, many fascinating applications have been in action due to the advent of NFC. Value chain is a concerted solvent. NFC mobile payment via NFC-micro SD application can be seen as the amended mobile payment resolution [4]. Sluggish acceptance of virtual point of sales (POS) terminals by merchants limit the potential of near-field communication (NFC) based payment devices. Mobile wallets promise to allow people to easily manage their accounts and to carry fewer cards [5]. NFC devices have been used either as device under attack or attack platform in several new attack incidents. The most prominent among them is the software-based relay attack [6]. To understand customers' perceived usefulness in exploring adoption models, a common approach has been followed and a goal-oriented paradigm is found to realize the perceived utility of the mobile wallet from a consumer's viewpoint [7].

4.4 TECHNOLOGY IN BANKING AND ACCEPTABILITY OF THE SAME BY THE STAKEHOLDERS

Updated computing system has empowered the individuals to be a part of the modern mode of digital transactions and has changed the dynamics of entire technological scenario. People have started accepting the new modes of transactions. With the incorporation of new chips in the plastic money, different cards have been catering multiple facilities to the users with convenience and efficacy. Similarly, virtual wallet has made the lives of users still easier as the users can use the same 24×7. Motto of Indian economy is to achieve optimal financial inclusion. However, to achieve this target, it's necessary to unfold why the users use these modes and why the non-users don't use them. In fact, highlighting the prime factors for users' motivation and non-users' demotivation is the need of the hour to make India digital. It is also important to highlight the usefulness of these modes for enhancing the chance of adoption of such gadgets. The study is primarily on customer satisfaction survey conducted in Bhubaneswar City. This survey was conducted in physical form. This result has been obtained by separating out the list of priority items and segregating the positive and negative impact on customers' adoption, usage frequency and satisfaction level. All the modes of digital transaction are embedded with high-end technology. Whatever excellent features may there be in the mode, it may not have the acceptability if the targeted customers have the fear for technology acceptance. Thus TAM theory is put in place for assessing the acceptability of the innovation in information technology (IT). Modified technical assessment model has been emphasized in TAM. Two distinct variables identified in this model are Perceived ease of use and Perceived usefulness (Figure 4.1).

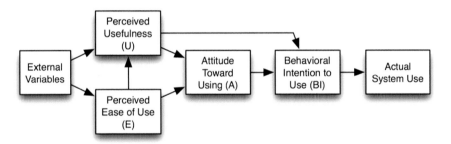

FIGURE 4.1   Technology acceptance model version one [8].

Primary data collection was done by interviewing a group of retail banking customers in Bhubaneswar City. Although the respondents could foresee the perceived usefulness of these digital modes of payments, most of them could not accept the same owing to negative perceived use of use. In fact, security aspect has been the main obstacle for the acceptance.

4.5   DANGER OF CYBER THREATS IN INDIA

India is now on the path of rapid growing high-end digital infrastructure as the head-count and approachability are speedily rising. However, digital push of India is yet to be accompanied with proper cybersecurity measures. As a result, many organizations are assailable to cyber-attacks. Common Cybersecurity threat incidents are website intrusions, phishing attacks, data damages, etc. The data of the Minister of State for Electronics reveals that in the first half of 2017, India has experienced more than 27,000 incidents of cybersecurity threat. As per the Indian computer emergency response team (CERT-In), the number of incidents concerning cybersecurity happened in recent years is presented in Table 4.1.

TABLE 4.1 Cybersecurity Incidents in India in Recent Years

Year	Cybersecurity Incidents
2014	44,679
2015	49,455
2016	50,362
2017 (till June)	27,482

Source: CERT-In.

The PwC and ASSOCHAM have done a collaborative study as per which there was an ATM card hack strike the Indian banks in October 2016. It affected about 3.2 million debit cards in the country. It is also revealed by the study that there have been about five times increase in Indian websites attacks in the past four years. While the cyber-attacks are in rising spree, budgetary provisions to address those are very dismal. India's budget allocation in 2012–2013 was only about Rs. 42.2 crore for cybersecurity purpose. As per the study, demonetization has given a push to e-wallet services, and mobile wallets have seen a gigantic outgrowth in downloads of Applications. In fact, there has been 100% growth in Application downloads for leading mobile wallets and wallet recharges have witnessed an epic 400% increase. Moreover, cyber threats will only rise as India is seeing a shift towards a cashless economy. The types of cybersecurity incidents such as phishing, scanning, website intrusions and defacements, virus code and DOS attacks will continue to grow," the study also added. It is a fact that most of the countries of the world are facing a dearth of specialists with the know-how, training, and drive required to fight out cyber criminals and India is no omission to it. Indian economy is subjected to ever-growing threat. Particularly, its major governmental departments, infrastructure set-up, and financial sector are highly vulnerable for cyber-attacks.

As per the advertisement of the enterprise security solutions brand of Quick Heal Technologies, there could be secret access to the servers and databases of more than 6000 organizations of India including Internet

Service Providers as well as Organizations from Public and Private sectors. It is reported that the hacker was offered 15 bitcoin (equivalent to approximately USD$73,000) for the same.

The list of recent incidents related to cybersecurity that badly affected India are:

- **Mirai Botnet Malware:** The malware targeted domestic users of router and other devices that are IoT-based. 2.5 million IoT devices have been affected by this malware.
- **WannaCry:** The cyber world was attacked by Ransomware WannaCry; May 2017. It was reported that banks in India were hit by this Ransomware. Moreover, it also affected badly some businesses in Tamil Nadu and Gujarat. It also affected Railware users.
- **Petya:** It is a Ransomware attack. India was affected by Petya ransomware attacks and nearly 20 organizations got affected by it.
- **Data Breaches:** 7.7 million users were stolen from Zomato by data breach. The stolen users were listed for sale on a Darknet market. Similarly, Data Breach also affected Reliance Jio.

In spite of threats of cybercrimes, some people have accepted different digital modes, data regarding which are mentioned in Tables 4.2 and 4.3.

In order to ensure that individuals and corporate are adopting digital modes of transactions in large scale, there must be efforts to strengthen cybersecurity.

4.6   DATA

On the basis of convenient sampling, 100 respondents, i.e., the retail banking customers based at Bhubaneswar, India (both users and non-users of digital modes-plastic money, net banking, and mobile wallets) have been studied and the broad data collected have been presented in Table 4.4.

4.7   FINDINGS

It has been found that retail banking customers, who don't use any of the modes of digital transaction, attribute the discredit of non-acceptance to their fear of getting cheated owing to security threats communicated to them through news capsules or words of mouth by their near and dear

TABLE 4.2 Transaction Volume and Value from Different Modes (Model 1)

Time Period	Real Time Gross Settlement		National Electronic Funds Transfer		Cheque Truncation System		Immediate Payment Service		National Automated Clearing House	
	Amount	Value	Amount	Value	Amount	Value	Amount	Value	Amount	value
Nov. 16	7.9	78479.2	123.0	8807.8	87.1	5419.2	36.2	324.8	152.5	606.6
Dec. 16	8.8	84096.5	166.3	11537.6	130.0	6811.9	52.8	431.9	198.7	626.8
Jan. 17	9.3	77486.1	164.2	11355.1	118.5	6618.4	62.4	491.2	158.7	541.4
Feb. 17	9.1	74218.8	148.2	10877.9	100.4	5993.9	59.7	482.2	150.5	592.0

Volume (in million) and Value (in Rs. Billion)

Source: RBI.

TABLE 4.3 Transaction Volume and Value from Different Modes (Model 2)

Time Period	Unified Payments Interface		Unstructured Supplementary Service Data		Point of Sale (Credit Cards and Debit Cards)		Prepaid Payment Instruments		Mobile Banking		Aggregate	
	Amount	Value	Volume in Thousand	Value (in Rs. Thousand)	Amount	Value	Amount	Value	Amount	Value	Amount	Value
Nov. 16	0.3	0.9	7	7302	205	352	59	13	72	1244	671	94004
Dec. 16	2.0	7.0	102	103718	311	522	87	21	70	1365	957	104055
Jan. 17	4.2	16.6	314	381760	265	481	87	21	64	1206	870	97011
Feb. 17	4.2	19.0	224	357055	212	391	78	18	56	1080	763	92594

Volume (in Million) and Value (in Rs. Billion)

Source: RBI (Decimals removed in few columns).

TABLE 4.4 Data of 100 Respondents

Gender	No. of Respondents	No. of Respondents Use Digital Mode of Transactions (Plastic Money, e–Banking, Virtual Wallets)				
		Debit/ATM Card	Credit Card	e-Banking	Virtual Wallets	Total
Male	60	42	4	7	2	55
Female	40	23	3	5	1	32
Total	100	65	7	12	3	87
Age Group	**No. of Respondents**					
Below 20	11	11	0	1	0	
20–30	18	15	3	2	2	
30–40	29	22	2	4	1	
40–50	19	12	2	3	0	
50–60	12	3	0	1	0	
60 and above	11	2	0	1	0	
Total	**100**	**65**	**7**	**12**	**3**	
Occupation	**No. of Respondents**					
Government Job	19	10	1	2	0	
Private Job	28	24	3	6	3	
Self-Employed	9	7	2	2	0	
Pensioners	12	2	0	0	0	
House-Wife	19	9	1	1	0	
Students	13	13	0	1	0	
Total	**100**	**65**	**7**	**12**	**3**	

Source: Primary data.

ones. However, those who use any or more of the modes have shown their mixed reaction regarding their experiences in using those. Table 4.5 represents the percentage of such users who use varied features inherent in one or other modes of digital transaction.

TABLE 4.5 Features of Digital Modes Used by the Respondents (Percentage of Users)

Features Used	Percentage of Respondents Who Use Any or More Digital Modes of Transaction
Direct access of website link	35
Blocking the card as phone gets stolen	66
Not having the photo-state copy of card	18
Memorizing the CCV number	35
Alerts regarding transaction	58
Password/Code of Transaction	65
Usage of Images	23
Using virtual card	15
Using virtual keyboard	11
Fixing the limit of the card	32

Source: Primary data.

Although a very small percentage of respondents have experienced monetary frauds during digital modes of transaction, almost all of them have the fear of losing money in next transaction. That's why most of them don't recommend any of their near and dear ones to adopt the same. Every user has the point to recommend to the government for strengthening the security aspect so that s/he will be confident while using and positive words of mouth may get aired. A small percentage of respondents who have been successfully using different modes have their points for the government and banks like educating the banking customers regarding the convenience of different modes and not deducting any commission or convenient charges for usage at least till the modes earn popularity.

4.8 CONCLUSION

Considering the findings of the study, it's inferred that sustainability of the Indian banking industry is in question. Customers' apprehension on

Financial Resolution and Deposit Insurance (FDRI) Bill-2017 has as such negated the confidence of customers on banks. Adding to that, exploitation of the customers by the banks and introduction and enhancement of charges for banking operations strengthen the question towards sustainability of banking industry and hence any dynamism in the form of new technology has been seen in suspicious mind of the customers.

Based on the analysis of experience and perception it is recommended that: (i) there must be creation of awareness among all the stakeholders including the customers about the availability and usefulness of modern gadgets of the banking sector; (ii) the customers need to be educated for safe usage of plastic money, net-banking, and virtual wallets so that their wrong perceptions can be wiped out; (iii) security aspect inherent in all those gadgets needs to be strengthened to safeguard the customers' money and develop sense of confidence among them pertaining to the usage of those gadgets, and most importantly, customers' satisfaction needs to be given due importance.

Need of the hour is to put concerted efforts for gross capacity building and set up updated and sophisticated cyber clinics that are competent of minutely examining all IT components before these are engaged in censorious base across the sectors of the country. In order to make the masses be in the cyberspace, budgetary allocation needs to be raised and trained personnel need to make the citizen aware of the benefits and how to use all such gadgets. Moreover, citizens also need to be trained for how to react when they encounter any cyber attack.

KEYWORDS

- banking industry
- changing banking norms
- competitive advantage
- customers' confidence
- cyber threats
- digital transaction
- retail banking customers
- sustainability
- technology adoption model

REFERENCES

1. Chandrasekhar, U., & Nandagopal, R., (2013). Mobile payments at retail point of sale: an indian perspective. *Life Science Journal, 10*(2), 2684–2688.
2. Thakur, R., & Srivastava, M., (2014). Adoption readiness, personal innovativeness, perceived risk and usage intention across customer groups for mobile payment services in India. *Internet Research, 24*(3), 369–392.
3. Halgaonkar, P. S., Jain, S., & Wadhai, V. M., (2013). A review of technology, tags, applications, and security. *International Journal of Research in Computer and Communication Technology, 2*(10), 979–987.
4. Wu, S. H., et al., (2013). Promoting collaborative mobile payment by using NFC-micro SD technology. *2013 IEEE International Conference on Services Computing*, Santa Clara, CA, 454–461, doi: 10.1109/SCC.2013.52.
5. Salajegheh, M., Priyantha, B., & Liu, J., (2013). *Taming the Wild Card for Mobile Payment*, MSR-TR-2013-61.
6. Michael, R., Josef, L., & Josef, S., (2013). Applying relay attacks to Google Wallet, *5th International Workshop on Near Field Communication (NFC)*, Zurich, 1–6.
7. Ho, D., Head, M., & Hassanein, K., (2013). Developing and validating a scale for perceived usefulness for the mobile wallet. In: Rocha Á., Correia A., Wilson T., & Stroetmann K. (eds) Advances in Information Systems and Technologies. *Advances in Intelligent Systems and Computing, vol 206*. Springer, Berlin, Heidelberg.
8. Davis, F. D., Bagozzi, R. P., & Warshaw, P. R. (1989). User Acceptance of Computer Technology: A Comparison of Two Theoretical Models, *Management Science 35*(8), 982–1003.

CHAPTER 5

e-Government Security Method Evaluation by Using G-AHP: A MCDM Model

PROSHIKSHYA MUKHERJEE, SASMITA RANI SAMANTA, and
PRASANT KUMAR PATTNAIK

KIIT Deemed to be University, Bhubaneswar, Odisha, India

ABSTRACT

Nowadays, e-government security is an important and crucial to trust maintain purpose among stakeholders to consume, process, and exchange the information over the e-government systems. Here, in this chapter, G-AHP (grey-analytic hierarchy process) method to help the policymakers to conduct a comprehensive assessment of e-government security strategy.

5.1 INTRODUCTION

e-Government is act as communication bridge between the government and citizen in transparent, efficient, and reliable manner through effective use of information technology (IT). Internet plays an important role for delivering the information from government to electronic document. Therefore, the security and privacy are most important for e-government applications [1]. The analytic hierarchy process (AHP) [2] is one of the powerful MCDM tools. However, effective evaluation is not available in AHP model. So, the grey-analytic hierarchy process (G-AHP) [3] is introduced for effectiveness of e-government security strategies evaluation purposes in multi-soft set approach.

The organization of the chapter as follows. In Section 5.2 discusses the literature review of e-government security. The basic idea of soft set theory is discussed in Section 5.3. Then the grey AHP is described and justified in Section 5.4. Finally, Section 5.6 discusses the conclusion and future scope of this research direction.

5.2   REVIEW OF E-GOVERNMENT SECURITY

For development of nation e-government is a key indicator as constructed by United Nation Public Administration Networks (UNPAN) [4]. Growth of e-government strongly depends on trust among citizen to process, store, and exchange the data through the e-government systems. The trust should be maintained by the effective security controls. That should be ensured that the unauthorized person cannot use the sensitive information.

In some research shows that the security issues are also affected in public management services [5, 6]. For e-government improvement trust plays an important role for effective and efficient transparent flow of data between citizens and government business [7]. Some developing countries less concerned about the security of e-government system [8]. So therefore, security is one of the factors for development of e-government system.

5.3   GREY ANALYTICAL HIERARCHY PROCESS

AHP method [2] is expressed as a complex decision making problem as a sequential set up hierarchy structure. Compute the comparatively weightiness measurement of diversified decision-making behaviors. In AHP for judgments matrix consideration 9 point scale (like extremely strong important, strongly important, moderately important, extremely unimportant, moderately unimportant, strongly unimportant, unimportant) is used. The steps of weight calculations are:

- **Step 1:** Pairwise comparison matrix calculation of the judgment matrix (M).
- **Step 2:** Relative weight (w_i) matrix calculation from the Eqn. (1).

$$w_i = G_i \bigg/ \sum_{n=1}^{N_F} G_n \qquad (1)$$

where, the G_i is the geometric mean of the ith row of the M judgment matrix is calculated as:

$$G_i = \sqrt[N_F]{a_{i1} a_{i2} \ldots a_{iN_F}} \tag{2}$$

where, N_F is the number of factors.

In GAHP [3] values are assigned on a scale of 9 to 1 scale against five grades: "very bad (1)," "bad (3)," "common (5)," "better (7)," "very good (9)" for framing the comment set of evaluation index $[v = (97531)^T]$. Intermediate grades between two adjacent grades are denoted with the following values: 8, 6, 4, and 2. The following steps are adopted for the calculation of the total effectiveness evaluation value:

- **Step 1:** Sample matrix (P) evaluation. The matrix is shown in Eqn. (3).

$$P = \begin{bmatrix} p_1^1 & \cdots & p_1^l \\ \vdots & \ddots & \vdots \\ p_{N_{SF}}^1 & \cdots & p_{N_{SF}}^l \end{bmatrix} \begin{matrix} Q_1 \\ \vdots \\ Q_{N_{SF}} \end{matrix} \tag{3}$$

where, p_m^n is the nth evaluation index value corresponding to sub-feature Q_m, l is the cardinality of the universal set U.

- **Step 2:** Grey cluster evaluation. Here, grey cluster divided in to the five grades: for "very good (g=1)," "better (g=2)," "common (g=3)," "bad (g=4)" and "very bad (g=5)."
- **Step 3:** Grey weight evaluation calculation shown in Eqn. (6).

Here, 1st we have to calculate the coefficient of grey cluster assessment. The Eqn. (4) shows the A_{lm}.

$$A_{lm} = \sum_{n=1}^{q} w_{\otimes l}(p_m^n) \tag{4}$$

Total quantity calculation with grey factor shows in Eqn. (5).

$$A_o = \sum_{p=1}^{N_{SF}} A_{po} \tag{5}$$

o^{th} alternatives grey weight calculation in Eqn. (6).

$$r_o = A_o \Big/ \sum_{o=1}^{|v|} A_o \tag{6}$$

Weighted matrix R_j shows in the Eqn. (7):

$$R_j = \begin{bmatrix} r_{j1}^1 & \cdots & r_{jq}^1 \\ \vdots & \ddots & \vdots \\ r_{j1}^{N_{SF}^j} & \cdots & r_{jq}^{N_{SF}^j} \end{bmatrix} \tag{7}$$

where, $j \in |a|$ where, a is non primitive matrix.

5.4   PROPOSED METHODOLOGY

In the proposed work have specially used grey AHP method for hierarchy forming. Figure 5.1 shows the four level hierarchy method, where, 1st level is goal then factor, sub-factor, and alternatives.

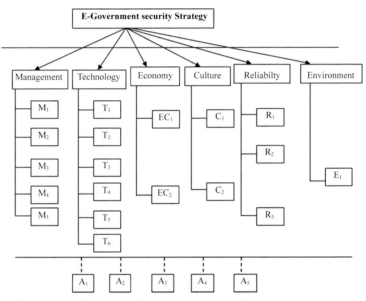

FIGURE 5.1   G-AHP framework for e-government security.

In the top of the figure, we defined e-government security policy our main goal to achieve. Subsequently, six criteria, like management (M), technology (T), economy (EC), culture (C), reliability (R) and environment (E) are listed in the figure.

In the 3rd level of the figure, every criterion has several sub-criteria. The sub-criteria are as follows; in management consist of IT policy availability (M_1), infrastructure availability (M_2), regular review (M_3), commitment (M_4), and standard (M_5). Technology consists of degree of cyber threats (T_1), new technology (T_2), national security threats (T_3), each point security (T_4), application security (T_5), and network security (T_6). Economy consists of attack cost (EC_1) and investment of security (EC_2). Culture consists of education security (C_1), and punishment and reward (C_2). Reliability consists of comfort (R_1), safety (R_2), and durability (R_3). The environment consists of production of environment (E_1).

In the bottom level shows the alternatives, here we considered five security objectives as the central concern in making any security decision. The alternatives are finance, health, defense, education, and agriculture environmental and climate.

5.5 MATHEMATICAL MODEL

In the proposed model, we assumed that under the subset $U = \{A_1, A_2, A_3, A_4, A_5, A_6\}$ a set of six alternatives and set of factors $E = \{M, T, EC, C, R, E\}$. Based on this the g-AHP model is discussed below:

- **Step 1:** In this step multi-valued table is constructed. Where, 'VG' stands for 'Very Good,' 'G' stands for 'Good,' 'M' stands for 'Medium' and 'B' stands for 'Bad.' Table 5.1 shows the multi-valued table.

TABLE 5.1 Multi-Valued Information System

Alternatives	Management	Technology	Economy	Culture	Reliability	Environment
A_1	G	G	G	G	G	G
A_2	M	M	G	M	G	M
A_3	M	G	B	M	G	G
A_4	B	B	M	M	B	M
A_5	VG	G	G	VG	VG	G
A_6	M	B	G	B	M	M

- **Step 2:** In this step pair wise comparison matrix is constructed and the Eqn. (8) shows the matrix. This matrix is generated by using AHP method.

$$
\begin{pmatrix}
1 & \dfrac{8}{7} & \dfrac{8}{6} & \dfrac{8}{5} & 1 & \dfrac{8}{9} \\[2mm]
\dfrac{7}{8} & 1 & \dfrac{7}{6} & \dfrac{7}{5} & \dfrac{7}{8} & \dfrac{7}{9} \\[2mm]
\dfrac{6}{8} & \dfrac{6}{7} & 1 & \dfrac{6}{5} & \dfrac{6}{8} & \dfrac{6}{9} \\[2mm]
\dfrac{5}{8} & \dfrac{5}{7} & \dfrac{5}{6} & 1 & \dfrac{5}{8} & \dfrac{5}{9} \\[2mm]
1 & \dfrac{8}{7} & \dfrac{8}{6} & \dfrac{8}{5} & 1 & \dfrac{8}{9} \\[2mm]
\dfrac{9}{8} & \dfrac{9}{7} & \dfrac{9}{6} & \dfrac{9}{5} & \dfrac{9}{8} & 1
\end{pmatrix}
\tag{8}
$$

- **Step 3:** In this step we have to calculate the relative weight from the Eqn. (1). The calculated weights are shown in below:

  $w_1 = 0.186$, $w_2 = 0.163$, $w_3 = 0.139$, $w_4 = 0.116$, $w_5 = 0.186$, $w_6 = 0.209$,

- **Step 4:** Table 5.2 shows the expert grade of sub-factor.

TABLE 5.2　Expert Grade for the Sub-Factor

	A_1	A_2	A_3	A_4	A_5	A_6
M_1	9	8	6	7	8	5
M_2	8	7	5	6	7	6
M_3	8	7	9	5	6	6
M_4	10	9	7	8	6	7
M_5	12	8	9	10	8	9
T_1	11	9	10	8	9	7
T_2	8	9	10	7	6	5
T_3	7	10	11	9	8	4
T_4	9	4	2	8	3	10
T_5	5	7	8	4	2	3
T_6	8	12	9	10	6	4
EC_1	7	5	4	5	7	6
EC_2	13	10	9	11	8	9
C_1	10	7	9	5	4	1
C_2	7	6	8	6	3	1
R_1	4	3	6	9	7	5
R_2	5	4	7	3	6	8
R_3	4	42	9	3	5	6
E_1	7	79	10	11	12	9

- **Step 5:** Grey assessment coefficient calculation by applying following Eqns. (9) to (13) and (4).

$$w_{\otimes 1}(A) = \begin{cases} A/9 & 0 < A < 9 \\ 1 & A \geq 9 \\ 0 & else \end{cases} \tag{9}$$

$$w_{\otimes 2}(A) = \begin{cases} 1 & 0 < A \leq 7 \\ \dfrac{14 - A}{7} & 7 < A \leq 14 \\ 0 & else \end{cases} \tag{10}$$

$$w_{\otimes 3}(A) = \begin{cases} 10 < A \le 3 \\ \dfrac{10-A}{5} 5 < A \le 10 \\ 0\,else \end{cases} \tag{11}$$

$$w_{\otimes 4}(A) = \begin{cases} 10 < A \le 6 \\ \dfrac{6-A}{3} 3 < A \le 6 \\ 0\,else \end{cases} \tag{12}$$

$$w_{\otimes 5}(A) = \begin{cases} 10 < A \le 1 \\ 1-A1 < A \le 2 \\ 0\,else \end{cases} \tag{13}$$

This coefficient belongs to the $w_{\otimes g}$ grey type is M_{1g}:

$$\text{If } g = 1, M_{11}(w_{\otimes 1}(A)) = w_{\otimes 1}(9) + w_{\otimes 1}(8) + w_{\otimes 1}(6) + w_{\otimes 1}(7) + w_{\otimes 1}(6) + w_{\otimes 1}(5)$$
$$= 1 + 0.889 + 0.667 + 0.778 + 0.556$$
$$= 4.779$$

In same way, to calculate, $g = 2$, $M_{12} = 5.428$, $g = 3$, $M_{13} = 3.4g = 4$, $M_{14} = 0.333g = 5$, $M_{15} = 0$.

The grey coefficient for alternative A_1 for the M_1 sub-factor is:

$$A1_1(A_1) = M_{11}(w_{\otimes 1}(9)) + M_{12}(w_{\otimes 2}(9)) + M_{13}(w_{\otimes 2}(9))$$
$$+ M_{14}(w_{\otimes 4}(9)) + M_{15}(w_{\otimes 5}(9))$$
$$= 1 + 0.714 + 0.2 + 0 + 0 = 1.914$$

In similar manner, $A_{11}(A_2) = 2.146$, $A_{11}(A_3) = 2.467$, $A_{11}(A_4) = 2.378$, $A_{11}(A_5) = 2.146$, $A_{11}(A_6) = 2.889$

Total assessment:

$$M_1 = A_{11}(A_1) + A_{11}(A_2) + A_{11}(A_3) + A_{11}(A_4) + A_{11}(A_5) + A_{11}(A_6)$$
$$= 1.914 + 2.146 + 2.476 + 2.378 + 2.146 + 2.889 = 13.94$$

Weights of the grey assessment vector for the subfactor M_1 are:

$$r_{11}^1 = \frac{A_{11}(A_1)}{M_1} = \frac{1.914}{13.94} = 0.137$$

$$r_{12}^1 = \frac{A_{11}(A_2)}{M_1} = \frac{2.146}{13.94} = 0.154$$

Same way, we find all vector weight.
- **Step 6:** The vector weights are listed in the matrix form in Eqns. (14)–(19).

Assessment weight vector calculation for management in Eqn. (14):

$$R_1 = \begin{bmatrix} 0.137 & 0.154 & 0.177 & 0.171 & 0.154 & 0.207 \\ 0.146 & 0.161 & 0.196 & 0.168 & 0.161 & 0.168 \\ 0.157 & 0.173 & 0.140 & 0.211 & 0.180 & 0.140 \\ 0.121 & 0.147 & 0.183 & 0.165 & 0.201 & 0.183 \\ 0.117 & 0.195 & 0.174 & 0.143 & 0.195 & 0.143 \end{bmatrix} \tag{14}$$

Assessment weight vector calculation for technology in Eqn. (15):

$$R_2 = \begin{bmatrix} 0.126 & 0.169 & 0.138 & 0.189 & 0.169 & 0.209 \\ 0.161 & 0.143 & 0.118 & 0.178 & 0.185 & 0.216 \\ 0.189 & 0.125 & 0.114 & 0.153 & 0.171 & 0.248 \\ 0.125 & 0.203 & 0.211 & 0.140 & 0.218 & 0.103 \\ 0.169 & 0.139 & 0.126 & 0.182 & 0.189 & 0.195 \\ 0.172 & 0.103 & 0.153 & 0.126 & 0.197 & 0.249 \end{bmatrix} \tag{15}$$

Assessment weight vector calculation for economy in Eqn. (16):

$$R_3 = \begin{bmatrix} 0.148 & 0.179 & 0.193 & 0.179 & 0.148 & 0.153 \\ 0.113 & 0.155 & 0.189 & 0.141 & 0.212 & 0.189 \end{bmatrix} \tag{16}$$

Assessment weight vector calculation for culture in Eqn. (17):

$$R_4 = \begin{bmatrix} 0.096 & 0.146 & 0.117 & 0.177 & 0.211 & 0.252 \\ 0.141 & 0.146 & 0.127 & 0.146 & 0.197 & 0.243 \end{bmatrix} \tag{17}$$

Assessment weight vector calculation for reliability in Eqn. (18):

$$R_5 = \begin{bmatrix} 0.200 & 0.214 & 0.159 & 0.123 & 0.153 & 0.151 \\ 0.167 & 0.159 & 0.155 & 0.218 & 0.161 & 0.140 \\ 0.187 & 0.194 & 0.115 & 0.181 & 0.174 & 0.149 \end{bmatrix} \tag{18}$$

Assessment weight vector calculation for environment in Eqn. (19):

$$R_6 = \begin{bmatrix} 0.227 & 0.182 & 0.150 & 0.136 & 0.123 & 0.182 \end{bmatrix} \tag{19}$$

- **Step 7:** The positive and negative ideal matrixes are listed in Eqns. (20) and (21):

$$PI = \begin{bmatrix} 0.157 & 0.195 & 0.196 & 0.211 & 0.201 & 0.207 \\ 0.189 & 0.209 & 0.211 & 0.189 & 0.218 & 0.249 \\ 0.148 & 0.179 & 0.193 & 0.179 & 0.212 & 0.189 \\ 0.141 & 0.146 & 0.127 & 0.177 & 0.211 & 0.252 \\ 0.200 & 0.214 & 0.159 & 0.218 & 0.174 & 0.151 \\ 0.227 & 0.182 & 0.150 & 0.136 & 0.123 & 0.182 \end{bmatrix} \quad (20)$$

$$NI = \begin{bmatrix} 0.117 & 0.147 & 0.140 & 0.143 & 0.154 & 0.140 \\ 0.125 & 0.103 & 0.114 & 0.126 & 0.169 & 0.103 \\ 0.113 & 0.155 & 0.189 & 0.141 & 0.148 & 0.153 \\ 0.096 & 0.146 & 0.117 & 0.146 & 0.197 & 0.243 \\ 0.167 & 0.159 & 0.115 & 0.123 & 0.153 & 0.140 \\ 0.227 & 0.182 & 0.150 & 0.136 & 0.123 & 0.182 \end{bmatrix} \quad (21)$$

Positive ideal solution (PIS) calculation:

$$\text{PIS} = \begin{bmatrix} w_1 & w_2 & w_3 & w_4 & w_5 & w_6 \end{bmatrix}^T [PI]$$
$$= [0.182\,0.189\,0.173\,0.184\,0.185\,0.201]$$

Negative ideals solution calculation:

$$\text{NIS} = \begin{bmatrix} w_1 & w_2 & w_3 & w_4 & w_5 & w_6 \end{bmatrix}^T [NI]$$
$$= [0.147\,0.150\,0.137\,0.135\,0.154\,0.156]$$

Relative closeness to the ideal alternative calculation:

$$A_1 = \frac{N_1^-}{P_1^+ + N_1^-} = \frac{0.147}{0.182 + 0.147} = 0.447$$

Similar way we have to calculate, $A_2 = 0.442$, $A_3 = 0.442$, $A_4 = 0.423$, $A_5 = 0.454$, $A_6 = 0.434$.

- **Step 8:** Ranking of the alternatives:

$$A_4 < A_6 < A_2 \leq A_3 < A_1 < A_5$$

The best alternative is A_5.

5.6 CONCLUSION

This chapter introduces Grey-AHP method for security strategy of e-government. The main feature is ability to capture vagueness and inconsistencies coming from subjective human judgments as decision-makers. This feature overcomes weakness on previous approaches based on classical AHP. Here, G-AHP method clearly explained and justified in real cases. In these days, the decision complexity and evaluation uncertainty is increased. This tool is used to solve this type of problem. The proposed decision-making model is not only robust and efficient but also more realistic and reasonable for real-world problem.

KEYWORDS

- **analytic hierarchy process**
- **e-government**
- **grey-analytic hierarchy process**
- **positive ideal solution**

REFERENCES

1. Jin-Fu, W., (2009). *e-Government Security Management: Key Factors and Counter-measure* (Vol. 2, pp. 483–486). IAS, 2009 5th International Conference on Information Assurance and Security.
2. Satty, T. L. *The Analytic Hierarchy Process: Planning, Priority Setting Resource Allocation*. McGraw-Hill, New York.
3. Jin, F., Liu, P., & Zhang, X. The evaluation study of knowledge management performance based on Grey-AHP method. *Eighth ACIS International Conference on Software Engineering, Artificial Intelligence, Networking and Parallel/Distributed Computing, IEEE Computer Society* (pp. 444–449).
4. Robin, H., (2010). *OECD Plans Creation New e-Government Indicators*. Future Government. http://www.futuregov.net/articles/2010/may/18// (accessed on 3 November 2020).
5. Backhouse, J., & Dhillon, G., (2001). Current directions in IS security research: Toward socio organizational perspectives. *Information Systems Journal, 11*(2), 127–153.
6. Dhillon, G., & Torkzadeh, G., (2006). Value focused assessment of information system security in organizations. *Information Systems Journal, 16*(3).
7. Richard, H., (2006). *Benchmarking e-Government: Improving the National and International Measurement*. Evaluation and Comparison of e-Government Published Research.
8. Hwang, J., & Syamsuddin, I., (2008). Failure of e-government implementation: A case study of South Sulawesi. *Proceeding of IEEE International Conference on Convergence and Hybrid Information Technology ICCIT2008* (pp. 952–960).

CHAPTER 6

e-Governance in the Health Sector

SUCHISMITA DAS, POONAM BISWAL, NISHTHA JAISWAL, and
SHARMISTHA BANERJEE

Kalinga Institute of Industrial Technology,
Deemed to be University, India

6.1 INTRODUCTION

In the 21st century, the world has turned into a global village of knowledge-based societies where the citizens are better informed and well connected. In the wake of these transformations there has been a paradigm shift towards electronic (e) governance or e-Governance.

The pillars of e-governance are: people; process; resources; and technology. The models used are:

- Government to government (G2G);
- Government to citizen (G2C);
- Government to employees (G2E);
- Government-to-business (G2B).

There is a wide range of scope for e-governance, like in the field of transportation, public safety, health and human services, education, and others. We will focus on the applications in the area related to healthcare. e-governance in health sector can be abbreviated as e-health or telehealthcare.

According to the EU, e-Health is an efficient way of accessing and the information and communication technologies (ICTs) across different field of application that affect the health sector, from the experienced specialist doctors to the hospital manager, via nurses, data processing specialists, social security of administrators and the patients. e-Health is basically delivering the healthcare to people outside the conventional care centers such as hospitals or residence; it can be explained as providing a patient with the means to alert a remote care provider of their need for assistance. It not only offers the ease of delivery, but also reaches out to the masses at their doorsteps in the remotest corners of the world by providing a single window opportunity which is efficient, rapid, transparent, and cost effective.

6.2   E-GOVERNANCE IN HEALTH SECTORS ACROSS THE GLOBE

There is a lot going around in the world for the well-being of the mankind, the major concern being the betterment of the health and living standards of people, this has made the countries to step out of their conventional methods and look for something that can reach to each and every individual, e-health plays a vital role in achieving this.

In Europe, The European Public Administration Network (EPAN) is focused on improving the quality and supply of information so that there is reduction in the cost, process time, and administrative burden which will lead to the improvement of service level and increase in the efficiency and customer satisfaction. The United States of America has adopted a strategy which is based on four areas: policy, tools, and methodologies, horizontal collaboration among countries, and knowledge management, the actions in these areas seek to strengthen health systems by integrating, decentralizing, and eliminating barriers in accessing the health services and also towards optimal management of infrastructure and human resources to promote community participation which will mobilize the support networks to establish cross-sector partnerships and public-private partnerships and to enhance the scientific and technological production and leverage the regional experience of public health programs. Likewise, the vision of the Swedish e-Health Agency is to lead, drive, and work in partnership for the development of e-Health and coordinate government initiatives by creating structures and monitoring quality that will manage electronic prescriptions, pharmaceutical information, and pharmaceutical statistics to provide different services for healthcare providers, pharmacies, and individuals.

6.3   E-GOVERNANCE IN THE HEALTH SECTOR IN INDIA

India is one of the most populated countries, with the highest diversity in people's biological characteristics. As varied is the geography of the country that varied is the different features found in people in different parts of the country. Even the body structures, types of diseases, health requirements differs according to the topography of the living areas, which makes the requirements of Indian people for better healthcare more heterogeneous and wide-ranging.

This led to the fame of using digital platforms for healthcare in the country. The ease and convenience of delivery of medical facilities at the doorsteps of people can increase the efficiency, rapidity, and transparency of medical services in a cost-effective manner. Corporate healthcare is increasing the use of the latest technology to provide best quality service. The Public health system, overburdened with increasing patients, is also opting towards the use of ICT in various parts of the country. Indian Government is also launching online communities for professional doctors and IT workers for the fusion of ICT and medical science. It reduces the delay, red tapism, chaos in big hospitals, rather encourages organizing, sharing, and accessing medical services.

6.3.1 CURRENT SCENARIO OF E-GOVERNANCE IN HEALTH SECTOR

Though all the above initiatives can be highly appreciated for good and efficient integration of ICT in providing medical services yet these are not enough to meet the needs of all people in India. Currently, e-Governance in India in healthcare is still unsatisfactory. Hundreds of Public Health Service Centers run by government are overburdened and collapsing. Large geographical size, increasing population density, lack of transport, inaccessibility, illiteracy, poverty, poor nutritional status and diversity in food habits and lifestyles make it difficult for e-systems to be implemented properly country-wide. Though metropolitan and many other smart cities across the country have shown an incredible shift to e-systems for efficient medical facilities yet there are many people who don't even get basic medical facilities in rural areas which arises a great concern.

6.3.2 HISTORY OF E-GOVERNANCE IN HEALTH SECTOR (E-HEALTH) IN INDIA

e-Governance was first initiated way back in 1987 with the launch of NICNET (National satellite-based Computer Network). Since then, the rapid development in IT sectors led to many initiatives being taken towards the implementations of e-health projects in India by both state and central government. In 2011, healthcare was added as a mission mode project (MMP) in the National e-Governance Plan (NeGP). It was also included in the government's Digital India Program launched in 2015 after the arrival of the new government. The government has launched three mobile health services namely, Kilkari, TB Missed Call, mCessation, and Mobile Academy for providing prior health services to the people and aid to medical professionals.

With the intention of taking medical-like service to the people door-steps, many other bold initiatives were taken. Some of them are as follows.

In Tamil Nadu, Tata Consultancy Services (TCS) has been allotted Rs 5 crore by government of Tamil Nadu for the development of an efficient solution to maintain electronic medical records (EMR). With the vision of functioning in 26 district headquarters, 162 talukas and 77 non-taluka hospitals, this web-based application to be created by TCS will allot each patient a unique ID. This ID will identify the entire data of previous medical records of the patient stored in web. The system is centralized, which makes it accessible from anywhere making it handy. Medical college hospitals in Tamil Nadu have applied ICT to manage in-patient and outpatient details, office automation, lab, and pharmacy services and medical records.

TNSACS (Tamil Nadu State AIDS Control Society) has applied web-based management for controlling 1100 VCTC, ART centers, Blood Banks, STD clinics, NGOs. Under the direct supervision of its Chennai head office, the number of HIV+ cases is kept under strict check.

HP heath care solution and Amrita Technology have achieved system integration and doctors' training via automation of 19 government hospitals and 14 medical colleges in Maharashtra. The JJ hospital (Grant Medical College) deals with 5-lakh OPD patient and 30,000 patients annually more efficiently due to the e-health and computerization resulted due to the above initiative of the companies.

Where in Delhi, Wipro has worked towards providing HIS (hospital information systems) to six hospitals. HIS has 28 modules (Patient registration, outpatients visit, doctors' appointments, demography, Admissions, Laboratory Results, etc.), fulfilling hospitals' needs. It has provided a large number of patients a better healthcare. Wipro has also initialized its idea of creating an electronic Patient Folder which would hold the details of each visit of individual patient in the six hospitals. This will help the doctors to access the details of patients easily, ensuring efficient patient treatment.

Government of Goa with 21st Century's Health Management Solutions has implemented a 2.5 crore Hospital Information System named Health NET in GMC (Goa Medical College). It aims toward improving the healthcare facilities of Goa by making it available for all the sectors of societies specially the poor, by increasing the efficiency and availability of health services. Patient Management Systems, Hospital Management Systems, the Laboratory management systems, Blood bank Management Systems, Advanced Imaging System Library and Academic Section Management System and Management Information Systems are some important provisions provided by the initiative.

In the state of Andhra Pradesh, CMC LTD has started the India Healthcare project which aims toward providing basic medical services to the people via PDAs. The primary health centers along with the auxiliary nurses and Midwives are being provided with PDAs so that during counseling they can collect data using it. All the data in PCs of public health centers can be transferred to these PDAs using available network. This application has already been implemented in 459 ANM in 67 PHC in Nalgonda district of Andhra Pradesh.

Odisha government has implemented three large projects for improving healthcare in the state. Tele Medicine and Tele referral services of NIC: The KBK districts (considered as backward districts of state) were connected via GRAMSAT network through which they were added into telereferral services of NIC. Many researches and conferences are being held every year on risk factors, diet, lifestyle modification, and prevention of CAD. A CME on "Prevention of sickle cell disease by DTN formula and genetic counseling" was organized by NIC Behrampur and conducted by doctors of all 30 districts of Odisha.

MKCG Hospital's telemedicine center is conducting monthly video conferencing with SGPGIMS on complex cases of different medical departments. Latest research on pulmonary scintigraphy in AIDS conducted at Nuclear medicine unit at MKCG hospital can be shared with the medical specialists via video conferencing.

e-Grama is an effort by NIC, Berhampur towards e-governance which main focus is to provide G2C (Government to Citizens Connection) to common man through different village level IT KIOSKS using ICT, which is an initiative and encouraging step from Dept. of IT, Government of

India. KIOSKS were opened by NGOs, youth clubs, villagers, and they access the internet portal from the centrally located NIC server through various remote access servers (RAS). It has implemented Odia and English language to provide static and dynamic services to the people. It has been successfully implemented in Ganjam district and other eight KBK districts. With the help of the IT kiosks, MKCG Hospital in Behrampur has undertaken a project to implement ICT tools for prevention of AIDS. Hello Doctor 24×7 is a mobile app which uses internet to provide medical facilities to remote areas of Behrampur. A project undertaken by final year students of MKCG college, Behrampur, whose aim is to enhance the healthcare system in rural area and to provide healthcare information via teleconsultation, telemedicine, specialist referral and emergency health-care. This initiative has been appreciated by DST, INTEL, and Indo-US Technology Forum. Around 80% of medical expenses can be saved by the people in rural area due to proper information and guidance provided by this project.

Intel has launched the World Ahead Programme as an initiative to provide proper education to health professionals in the country. Intel has done tele-health projects in Maharashtra, Baramati, and Trivandrum. Rui hospital in collaboration with Aurobindo Eye Hospital Madurai and Narayan Hrudayalaya Bangalore has started tele-health services for eye and heart patients. Intel has also established a health monitoring system in St. Philomena Girls' Higher Secondary School, Trivandrum with the help of TCS.

6.4   NEED OF ICT IN HEALTH SECTOR

In 2011, as per census, India's population is at 1.21 billion (0.62 billion males and 0.588 females) but healthcare system and facility is not equally maintained in each and every state of India. The infant mortality in Kerala is 6 per 1000 live births, whereas in Uttar Pradesh, it is 64. Malnutrition among rural people being the major issue, Malaria, Dengue, Typhoid seems to remain predominant throughout the year all over the country. These circumstances create huge demand for healthcare service to upgrade themselves so that they can reach out to all in need within affordable time and cost. The demands of 1.21 billion cannot be fulfilled by 1.8 million health professionals physically, so ICT's involvement is required to improve the scenario. It can be beneficial in the following ways:

1. It is simple. It can be easily handled by patients, providing them basic healthcare information and assistance from anywhere.
2. It can easily reach the remote parts of the country where the medical service is very poor and the health status of the people is deteriorating. In the case of mild injuries or sickness, it can save the patients' time and money from traveling to the cities for treatment. These platforms can also assist them by consulting them to the specialists and hospitals nearby and can also provide them appointments via online media.
3. It is highly cost-effective, can save about 80% of medical expenses of the patient, and also saves the run to the hospitals.
4. It can provide basic care to the patient in case of emergency or delay of medical assistance.
5. The Local health centers can remain in contact with the city hospitals can take help and advises from the professionals there. When required, can also send patients to the hospitals by giving prior appointments.
6. These platforms can act as the greatest and most efficient forum for professionals to share their knowledge, research, and cases with the learning students of different medical colleges. It can improve the quality and ability of the next generation doctors.

6.5   FUTURE SCOPE OF E-GOVERNANCE IN INDIA

The efficient application of ICT in India can change the health sector tremendously. It can make it efficient, time-saving and cheap. India, where still thousands of people have below-average income, can easily afford the health services which they can't now. Secondly, people living in the remote areas, mountainous regions, and plateaus find it difficult to access the nearest hospitals which may be hundreds of kilometers away from them, ICT, and e-Systems can resolve this by giving them medical services at their doorsteps. ICT can also act as a learning platform for medical students, it can act as a medium via which more advanced, skilled doctors can be generated, that too more in numbers as doctor to patient ratio in

India is still very low. It can encourage the use of health apps which aids in increasing the awareness among the people about health. It can also act as a secure platform to avail health insurances, birth certificates, death certificates. All in all, ICT can save millions of lives in the country who die either due to lack of medical care or time.

6.6 CONCLUSION

Though many initiatives have been taken for proper functioning of ICT in India yet they are not enough to meet the needs of billion Indians. More large, advanced, and integrator projects regarding the usage of e-systems should be made in order to reach each corner of the country and providing the people with efficient basic healthcare services. This overview of e-governance is aimed towards increasing the awareness of people towards ICT use in health sector, to evoke interest in medical students in learning ICT and save precious time of both patients and doctors. It encourages the use of e-systems in order to provide quality healthcare services at the doorsteps from anywhere and anytime.

KEYWORDS

- **electronic medical records**
- **European Public Administration Network**
- **mission mode project**
- **National e-Governance Plan**
- **remote access servers**
- **Tata consultancy services**

REFERENCES

1. Mahapatra, S. C., Das, R. K., & Patra, M. R., (2011). *Current e-Governance Scenario in Healthcare Sector of India*. CSI Publication.

2. Gole, I., Sharma, T., & Misra, S. B., (2017). Role of ICT in healthcare sector: An empirical study of Pune City. *Journal of Management and Public Policy, 8*, 23–320. New Delhi.

3. Bhatnagar, S., (2014). *Public Service Delivery: Role of Information and Communication Technology in Improving Governance and Development Impact.* ABD Economics Working Paper Series, Asian Development Bank.

4. Das, R. K., & Dash, S. S., (2007). Telereferal service of NIC: A helping hand for the doctors and inhabitants of KBK districts of Orissa. In: *Book: Adopting e-Governance* (pp. 253–258). GIFT Publishing.

5. Lele, R. D., (2008). *ICT in Day-to-Day Clinical Practice Postgraduate Medicine API and ICP, XXII,* 3–9.

CHAPTER 7

Vulnerability and Cybersecurity in the Natural Gas and Oil Industry

SUBHAM NASKAR,[1] PATEL DHRUV,[1] PRANJAL KUMAR,[1] and
SOUMYA MUKHERJEE[2]

[1]*Kalinga Institute of Industrial Technology, Bhubaneswar, Odisha, India*

[2]*Government College of Engineering and Ceramic Technology, Kolkata, West Bengal, India*

7.1 INTRODUCTION

With an estimated annual turnover of around 87.5 trillion dollars and being a provider of 74% of the energy around the globe it is really important for the industrialists and action managers of the energy industries to focus on cybersecurity, in order to bolster the digital monitoring of the whole system. Any type of impact can incur energy disruptions and extreme losses around the globe. Therefore, to prevent any of such technical failures or any other digital related breaches the company employees must ensure the protection of operational technologies and industrial control systems [1].

7.2 CRITICAL STEPS OF CYBERSECURITY IN AN ENERGY PRODUCTION UNIT

1. **Network Security:** It determines the measures to be taken for shielding a communication pathway from unapproved or accidental access to normal operations [15].
2. **Identity and Access Management:** The part of cybersecurity that empowers the required people to receive correct assets at correct occasions for correct reasons [2]. IAM addresses the strategic need

to guarantee proper access to assets over progressively heterogeneous innovations, and to progressively meet the interior necessities thoroughly. This security practice is a critical endeavor for the common gas and oil industry. It is progressively business-adjusted, and it requires business abilities, along with specialized expertise.

3. **Data Protection:** Securing advanced information, for example, those in a database, from unethical reach maliciously done by unapproved clients. Data security protects user-facing regions from possible data breaches.

4. **Application Security:** It helps in shielding an application or site from assault, including constant analysis of the web and portable applications just as well as web application firewalls.

5. **Endpoint Security:** The way toward verifying the different endpoints on a system including the cell phones, workstations, and work areas, servers in an information focus, and also estimating the dangers exhibited by gadgets interfacing with a venture network. The significance of this part is increasing progressively with more prominent utilization of cell phones, and the endpoint security ensures protection for the corporate system allowing the endpoint gadget to work outside of the system utilizing the cloud facilities without undermining the security measures.

6. **Vulnerability Management:** The practice of distinguishing, grouping, remediating, and moderating vulnerabilities, especially in programming.

7. **Threat Protection:** It falls into the category of cybersecurity arrangements that provide protection against malware or hacking-based assaults occurring mainly on vulnerable data.

8. **Risk and Compliance:** The examination of outside and inner possibility for flaws that can disclose the resources, and the execution of viable inner arrangements for preventing breaches this step can provide cybersecurity measures for an organization.

9. **Forensics and Insider Risk:** Digital legal sciences incorporate the recuperation and examination of flagged items found in computerized devices. This part of the administration incorporates exercises [2]. For example, client conduct examination as well as endpoint checking proposed to distinguish potential malevolent exercises by a present or previous worker, contractual worker or other individuals who has access or had approved access to an association's frameworks arrangement, or internal undisclosable info [15].

7.3   GOVERNMENT AND REGULATORY COOPERATORS

1.   **Transportation Security Administration (TSA):** Government endeavors identified with pipeline are secured by the Office of Security Policy as determined by TSA. With the help of industry as well as government individuals from the Pipeline Sector and the Government Planning Councils, industry affiliation delegates, and other invested individuals, TSA built up the Pipeline Security Guidelines. Using a comparative industry and government communitarian approach, these rules are consistently refreshed to mirror the progression of security practices to meet the consistently evolving risks in both the physical and cybersecurity realms. Gaseous petrol and oil organizations contributed to TSA as it created and refreshed the Pipeline Security Guidelines. Pipeline administrators joined forces with TSA through its Pipeline Corporate Security Review program as TSA finished audits of all the country's main 100-pipeline frameworks, which transport 84% of the country's energy.

2.   **Department of Homeland Security (DHS):** It drives the Federal government's endeavors to verify the country's basic framework by working with proprietors and administrators to get ready for, avoid, relieve, and react to risks. Along with industry, the DHS Office of Framework Protection (IP) division of the National Protection and Programs Directorate (NPPD) directs and organizes national projects. The workplace conducts out evaluations to help basic foundation proprietors, the administrators, and the state. Regional accomplices comprehend and address dangers to basic foundation. IP gives data on developing dangers and risks with the goal and the proper moves that should be conducted.

3.   **U.S. Coast Guard (USGC):** Cybersecurity specialists from petroleum gas and oil organizations have worked cooperatively with the USCG and their counselors from the NIST National Cybersecurity Center of Excellence (NCCOE) also, from The Miter Corporation to build up the required security protocols. Together, these specialists co-characterized the strategic destinations of flammable gas and oil offices and tasks and characterized the parts of the NIST CSF that ought to be underscored by organizations to moderate the devastation that can be caused by a digital assault.

4. **Oil and Natural Gas Subsector Coordinating Council (ONG SCC) and Energy Sector Government Coordinating Council (EGCC):** The business is drawn in with the ONG SCC and EGCC, data sharing bodies that cut crosswise over for all intents of every government organization included in cybersecurity-identified with the flammable gas and oil industry. The ONG SCC gives a setting for industry proprietors and administrators to examine area wide security projects, systems, and forms, trade data and survey achievements and progress toward consistent improvement in the assurance of the area's basic foundation. The EGCC gives a private gathering for powerful coordination of security methodologies just as exercises, strategies, and correspondence over the area to help the country security. The EGCC tries to fill in as a solitary purpose of contact to encourage correspondence between the administration and the private segment when getting ready for and reacting to issues and dangers coming about because of physical, digital or climate-related events affecting the vitality part.

7.4   WHY IS THIS INDUSTRY SO VULNERABLE?

The vital energy business will be kept under high-risk analysis since the slightest flaw can result in devastation, especially its significance should be almost equal to national and monetary security. We expect that the accompanying circumstances may further add to the risk factor in this business [4]:

1. With the steep drop in oil costs, the APT gatherings' and numerical analytics in the estimation of stolen oil or gas might change drastically.
2. Continued advancements in petroleum and the derivative improvement will most likely be expanded and the digital secret activities as the APT gatherings might be attempted to be stolen for extracting intellectual property from it by rivals or any unethical personalities.
3. The growing interest for this branch worldwide will probably bring about expanded digital secret activities and cybercrime for extracting such info will also show a steep rise.
4. Observed undercover work by presumed Russian-based risk gatherings leading surveillance of mechanical control frameworks (ICS) and supervisory control and information procurement (SCADA) frameworks.

5. The conflict between nations could likewise bring about expanded danger action as state-supported risk entertainers may look to increase the pressure on a rival by upsetting their vital supply.
6. Environmental issues and different debates identified with vitality generation may likewise bring about expanded risk action from hacktivists looking to point out the issues and humiliate associations that they claim to be responsible for [4].

7.5 PUBLIC POLICIES FOR THESE ORGANIZATIONS

7.5.1 *CYBERSECURITY ACT OF 2015*

Expects organizations to secure data what's more, share as per certain conventions [10]:

1. Provides lawful securities to organizations at the point when these necessities are met.
2. Establishes DHS as a center point for data sharing, giving a channel to digital risk markers to go back and forth the private part to the U.S. Government, along with other insight organizations.
3. Incentivizes the work of both Information Sharing and Analysis Centers (for example, ONG-ISAC; the Oil and DNG-ISAC, Natural Gas ISAC and Downstream Natural Gas ISAC).

7.5.2 *EXECUTIVE ORDER 13636*

1. Develop a nonpartisan voluntary cybersecurity structure.
2. Promote and boost the selection of cybersecurity procedures.
3. Increase the volume, practicality, and quality of digital risk data sharing.
4. Incorporate solid security and common freedoms assurances into each activity to secure the basic foundation.

7.5.3 *PRESIDENTIAL POLICY DIRECTIVE-21*

Build up a situational mindset that includes both physical and digital parts of how the process is working side by side:

1. Understand the outcomes of the infrastructure failure;
2. Evaluate utilize the public and private association;
3. The National Infrastructure Security Plan must be updated;
4. Extensive research and improvement plan should be developed.

7.6 THREAT ANALYSIS BY CERTAIN ORGANIZATIONS

Dragos, at Black Hat 2019 [3], recognized another danger bunch fit for ICS assaults that are as of now focusing on oil and gas enterprises. Named "Hexane," the gathering has been active since 2018 and is likewise focusing on communication organizations in the Middle East, Central Asia, and Africa. Dragos specialists trust Hexane could be intending to utilize telecom suppliers "potentially as a stepping stone to network-focused man-in-the-middle and related attack."

The rise in ICS security dangers against oil and gas organizations has concurred with a spike in political clashes between different countries over the globe. "In the course of the most recent year and a half, as pressures have ascended far and wide, oil, and gas has become a prevalent objective," Sergio Caltagirone, VP of risk knowledge at Dragos, told Search Security [13].

Dragos' report noticed that just a single risk gathering, Xenotime, has exhibited an ability to not just infiltrate ICS arranges in oil and gas offices but could also possibly cause a dangerous occasion. Caltagirone said the potential for a cyberattack to prompt death toll is higher at oil and gas offices than electric utilities in light of the fact that the unstable treatment facility procedure includes combustible materials and might prompt blasts [8].

7.6.1 GROUPS TARGETING OIL AND GAS

1. **Xenotime [5]:** It caused a breakthrough at an oil and gas office in the Kingdom of Saudi Arabia in August 2017 utilizing the damaging TRISIS system, uniquely custom fitted to communicate with Triconex wellbeing controllers. The TRISIS assault spoke to a heightening of ICS assaults because of its potential disastrous capacities and outcomes. In 2018 Xenotime action extended to incorporate oil and gas organizations in Europe, the US, Australia,

and the Middle East; electric utilities in North America and the APAC area; just as gadgets past the Triconex controllers. This gathering additionally undermined a few ICS merchants and makers, giving a potential production network threat.

2. **Magnallium [7]:** It has focused on petrochemical and aviation producers since in any event 2013. The movement bunch at first focused on an airship holding organization and vitality firms situated in Saudi Arabia; however, it extended their focusing to incorporate substances in Europe and North America. Magnallium's capacities seem to at present do not have ICS-explicit ability, and the gathering stays concentrated on introductory IT intrusions.

3. **Chrysene:** It created from an undercover work battle that initially picked up considerably after the dangerous Shamoon cyberattack in 2012 that affected Saudi Aramco [9]. The movement gathering targets petrochemical, oil, and gas, and electric age divisions. Focusing on has moved past the gathering's underlying spotlight on the Gulf Region and the gathering stays dynamic and advancing in more than one territory [10].

4. **Hexane:** It targets oil and gas and broadcast communications in Africa, the Middle East, and Southwest Asia. Dragos distinguished the gathering in May 2019. Dragos can just freely share restricted data about this recently distinguished movement bunch at this time [11].

5. **Dymalloy:** It is an exceptionally forceful group that can accomplish long haul and determined access to IT and operational conditions for insight accumulation and conceivable future interruption occasions. The gathering's unfortunate casualties incorporate electric utilities, oil, and gas, and propelled industry elements in Turkey, Europe, and North America.

In the present risk scene, no enemy has exhibited the purpose, inspiration, or capacities to target midstream ICS/PCN conditions. The enemy doubtlessly ready to build up these capabilities is Xenotime. Security penetrators have directed digital tasks focusing on transport and business activities at oil and gas substances to an obscure degree-for example, in April 2018, aggressors focused on electronic information exchanges (EDI) at various US vitality organizations making disturbances business activities [14].

7.7 PROBABLE APPROACH FOR DEFENSE AGAINST ATTACKERS

1. **Visibility:** A far-reaching approach for permeability into ICS/ OT situations ought to be taken to guarantee that there isn't a permeability hole. Resource proprietors and security staff should cooperate to assemble the system and host-based logs beginning from the most basic framework. The capacity to distinguish and connect suspicious system, have, and operational occasions can incredibly aid either distinguishing interruptions as they happen, or encourage underlying driver investigation after a problematic occasion. Guarantee organizes observing of the ICS through ICS-centered advances [6].

2. **Segment:** Where conceivable, fragment, and separate systems to restrain foe horizontal development abilities. While physically troublesome in existing situations, present-day organizing equip-ment may empower resource proprietors and administrators to for all intents and purposes fragment systems to lessen assault surface and cut off the assaulters.

3. **Accessibility:** Identify and sort entrance and departure courses into control framework systems. This incorporates architect and manager remote access entrances, yet additionally covers things, for example, business knowledge and authorizing server to connect that needs to get to IT assets or access more extensive part of the web. Farthest point these sorts of associations, including firewall rule directionality, to guarantee a limited uncovered assault area.

4. **Public Data:** Assess resource proprietor facilitated, freely posted data and information that, when totaled, would create sensitive data that could be used by an attacker. Work with merchants, contractual workers, and different parties-either casually or through formal necessities in contracts-to limit or avert recognizable proof of explicit locales, abilities, or gear in advertising or related material [6].

5. **Configuration:** Identify and store "known great" design data for ICS gadgets in nonnetwork available areas to give baselines to the examination just as reestablish focuses on the occasion of disturbance. Update these things as often as possible to guarantee such stockpiling mirrors creation conditions [16].

   This activity not just helps recuperation in case of IT malware engendering into ICS systems, yet in addition encourages examination in TRISIS-like occasions by giving baselines to think about possibly the controlled designs.

6. **Defense-in-Depth:** Design and execute safeguard inside and out encompassing ICS systems where security controls and improved permeability are applied to be capable of taking care of such undertakings. Models incorporate requiring remote access to the course through a hop host including improved Windows and system logging to guarantee sufficient checking of remote access to the control framework arrange.

7. **Consequence-Driven:** Identify and organize basic resources and associations, and procedure outcomes of cyberattacks.

8. **Third-Parties:** Ensure that outsider associations and ICS cooperation is observed and logged, from a "Trust, yet Verify" mentality. Where conceivable, separate or make unmistakable enclaves to guarantee that outsider access doesn't bring about complete, liberated, or unmonitored access to the whole ICS arrangement.

9. **Network Infrastructure:** Allanite and Dymalloy [12] routinely target switches during bargains, changing setups to take into consideration diligent access or conveyance of extra malware. Execute a firewall arrangement audit to guarantee attackers don't mess with designs and find security holes.

10. **Response Plans:** Develop, survey, and practice digital assault reaction designs and incorporate digital examinations concerning main driver investigation for all occasions.

7.8   INTEGRITY THAT SHOULD BE ESTABLISHED WITHIN THE ORGANIZATION AS DETERMINED FROM A SURVEY

1. **The Awareness of the Worker is the Most Important Part:** 78% think about an indiscreet individual from staff as the probably wellspring of an assault. 43% of noteworthy digital ruptures were from an absence of end client mindfulness, abused through phishing [8].

2. **Data Security Needs Board-Level Consideration:** 87% have not completely considered the data security ramifications of their present procedure and plans. 46% feel the entire board is educated about data security [16].

3. **The Hazard to Notoriety is Rising:** 60% have had an ongoing huge cybersecurity episode.

    15% have a powerful episode reaction program and normally direct tabletop works out [17].

4. A digital cybersecurity workforce is expected to keep up the pace with advancing dangers.

    Half of the participant state the absence of gifted valuable assets has a negative impact on the association. 95% state their cybersecurity capacity doesn't completely address their association's issues.

5. **Difficulties are on the Ascent with the Internet of Things (IoT):** 17% feel almost certainly, they would distinguish a refined digital assault.

6. **The Monetary Effect of Ruptures isn't Completely Inspected:** 97% of the associations' data security reports don't calculate money related effect of each breach.

Around 63% would not expand their cybersecurity spending in the wake of encountering a break that didn't seem to do any mischief.

KEYWORDS

- **cybersecurity measures**
- **cybersecurity protocols**
- **internet of things**
- **natural gas**
- **oil industries**
- **vulnerability**

REFERENCES

1. https://www.thesslstore.com/blog/80-eye-opening-cyber-security-statistics-for-2019/ (accessed on 3 November 2020).
2. https://www.accenture.com/us-en/insights/security/cost-cybercrime-study (accessed on 3 November 2020).

3. https://dragos.com/wp-content/uploads/Dragos-Oil-and-Gas-Threat-Perspective-2019. pdf (accessed on 3 November 2020).

4. https://theonebrief.com/terrorism-political-violence-risk-impact-to-oil-energy-industry/ (accessed on 3 November 2020).

5. https://dragos.com/resource/xenotime/ (accessed on 3 November 2020).

6. https://attack.mitre.org/groups/G0088/ (accessed on 3 November 2020).

7. https://dragos.com/resource/magnallium/ (accessed on 3 November 2020).

8. https://attack.mitre.org/groups/G0064/ (accessed on 3 November 2020).

9. https://dragos.com/resource/chrysene/ (accessed on 3 November 2020).

10. https://attack.mitre.org/groups/G0049/ (accessed on 3 November 2020).

11. https://dragos.com/resource/hexane/ (accessed on 3 November 2020).

12. https://dragos.com/resource/dymalloy/ (accessed on 3 November 2020).

13. https://attack.mitre.org/groups/G0074/ (accessed on 3 November 2020).

14. https://dragos.com/blog/industry-news/threat-proliferation-in-ics-cybersecurity-xeno-time-now-targeting-electric-sector-in-addition-to-oil-and-gas/ (accessed on 3 November 2020).

15. https://www.ey.com/en_gl/oil-gas/six-cybersecurity-issues-for-oil-and-gas-companies (accessed on 3 November 2020).

16. https://www.fireeye.com/content/dam/fireeye-www/current-threats/pdfs/ib-energy. pdf (accessed on 3 November 2020).

17. https://www.fireeye.com/current-threats/reports-by-industry/energy-threat-intelligence.html (accessed on 3 November 2020).

CHAPTER 8

An Encryption Scheme: ECC and Arnold's Transformation

JIBENDU KUMAR MANTRI,[1] RAJALAXMI MISHRA,[2] and
PRASANTA KUMAR SWAIN[1]

[1]Department of Computer Application, North Orissa University, Odisha, India

[2]College of IT and Management Education, Bhubaneswar, Odisha, India

ABSTRACT

Messages communicated through the network are vulnerable to unauthorized access. The confidential messages are required to be protected from unauthorized access. The most commonly used form of messages is text and images. Elliptic curve cryptosystem provides more security compared to other cryptosystems. Researchers have proposed many cryptosystems based on elliptic curve cryptography for text as well as for images. We come up with an encryption scheme using Arnold's transformation and ElGamal encryption scheme based on elliptic curve cryptography in this chapter. The result of the implementation and scrutiny of security features show the strength of the proposed scheme.

8.1 INTRODUCTION

In 1985, Victor Miller and Neal Koblitz proposed separately the utility of an elliptic curve in cryptography. They established the fact that for a comparatively smaller key size, elliptic curve cryptography offers better security. Their security scheme was based on the hardness of a different problem, the elliptic curve discrete logarithmic problem (ECDLP). Many researchers have used Elliptic curve cryptosystem and confirmed that it is more secure than other cryptosystems.

V. I. Arnold proposed Arnold transform in the research of ergodic theory, also called cat mapping, which is applied to digital images. In this research work, we have used Arnold's cat map to scramble an image, then the scrambled image is encrypted using the ElGamal scheme based on Elliptic curve cryptography. The rest of the chapter is arranged as follows: Section 8.2 includes different related literatures, Section 8.3 summarizes the basics of elliptic curve cryptography and Arnold's transformation, our proposed algorithm is presented in Section 8.4, and Section 8.5 contains simulation outcomes, and in Section 8.6 the proposed scheme is analyzed according to security point of view, and finally, it contains the concluding remarks and future direction in Section 8.7.

8.2   LITERATURE SURVEY

The utility of Elliptic curves in cryptography was explained by Victor S. Miller [16]. He described a scheme for faster encryption scheme analogous to the Diffie-Hellman key exchange scheme. Neal Koblitz [11] described the asymmetric key or public key cryptosystem using elliptic curves over finite fields. He clarified that the discrete logarithm problem (DLP) used in finite group field is harder than DLP over binary field. Neal Koblitz, Alfred Menezes, and Scott Vanstone [12] used elliptic curve group and used the idea of discrete logarithmic problem. It provided a smaller block size, high speed and more security. Darrel Hankerson, Alfred Menezes, and Vanstone [5] discussed the different arithmetic of elliptic curve, issues of implementation, and several cryptographic protocols in detail. Lawrence C. Washington [9] verified various theories associated to elliptic curve. An encryption scheme using chaotic system and cyclic elliptic curve for images was presented by Ahmed A, Abd El-Latif, and Xiamu Niu [1]. They explained a technique of encryption for images where a chaotic system and cyclic elliptic curve point were used for generation of a pseudo-random key stream. A cryptanalysis of encryption technique for images was presented by Hong Liu and Yanbing Liu [7]. Their work is based on cyclic elliptic curve with hybrid chaotic system. S. Maria Celestin Vigila and K. Muneeswaran [15] presented a technique of encryption for images using ECC. They generated, randomly the private key and an integer 'k' by using a coupled linear congruential generator of random numbers. To obtain the ciphered image they used the point multiplication operation of the generator point of the curve with each pixel value to map into the elliptic curve coordinate. A mapping table is required to decipher the image. Ali Soleymani, Md Jan Nordin, and Zulkarnain Md Ali [2] described a scheme of encryption by making use of elliptic curve over prime field. They used a mapping table having row indexes as the values from 0 to 255 and the content in the given row is one point on elliptic curve. The table is used to map the intensity of the pixel values onto the points on elliptic curve. The receiver's public key is used to encrypt the plain image. To get the ciphered image the mapping table is again used to associate the points with the values in the range of 0 to 255. Behnia et al. in their work [14] explained an image encryption scheme based on the Jacobian elliptic map. They transformed the data matrix of plain image into a one-dimensional matrix by doing the operation with the key. The elements of the matrix are enciphered by using an equation and the matrix is converted to the original dimension. The homomorphic scheme of encryption for sharing secret images using ElGamal based on elliptic curve was proposed by Li, Ahmed A. Abd El-Latif, Xiamu Niu [10]. To prevent Pollard's rho attack, isomorphism, and Pohlig Hellman attack they selected the elliptic curve parameters. The experimental outcomes reflect that it is a better encryption technique than encryption schemes using ElGamal and RSA.

Don Johnson, Alfred Menezes, and Scott Vanstone [6] explained various issues regarding the implementation and security issues of the Elliptic Curve Digital Signature scheme. Ann Hibner Koblitz, Neal Koblitz, Alfred Menezes [3] presented the twists and turns of the progression of ECC and gaining approval of people with time. W. Stalling [17] explained in detail the network security issues and elaborated different cryptographic algorithms. Singh and Singh [8] explained the scheme for image encryption/decryption and enclosure of digital signature to the encrypted image to provide genuineness and integrity. They have grouped the pixel values, then the grouped pixel values can be paired to represent a point on the Elliptic curve. Omar Reyad [13] presented a scheme of encoding the plaintext message and then applying elliptic curve operations. He used a mapping method to encode the ASCII value to a point on the elliptic curve. For decoding he used reverse of mapping method. Binamy Kumar Singh, Abusha Tsegaye, and Jagat Singh [4] proposed a probabilistic method of encryption using elliptic curve and Arnold transformation.

8.3   BASICS OF ELLIPTIC CURVE CRYPTOGRAPHY

The elliptic curves are different from ellipses. They are represented by the cubic equations similar to the equations used to find out the circumference of the ellipse. The elliptic curve operations defined over real numbers are slow and flawed due to round-off errors. The operations used in cryptography require being efficient and precise. To make operations on elliptic curve accurate and faster, the curve cryptography is defined over two finite fields:

- Prime field Fp; and
- Binary field $F_2{}^m$.

For cryptographic operations, the field chosen with finitely large number of points is suitable. The elliptic curve operations are defined on affine coordinate system, where each point is represented by the pair (x, y).

8.3.1   ELLIPTIC CURVE CRYPTOGRAPHY OVER PRIME FIELD FP

Let there is a prime number p>3. Then E is the elliptic curve over Fp defined by the equation called Weierstrass equation which is of the form:

$$y^2 \bmod p = (x^3 + ax + b) \bmod p \qquad (1)$$

where, a and b are coefficients of the equation of elliptic curve belong to prime field Fp, and they satisfy the non-singularity condition for the elliptic curve:

$$(4a^3 + 27b^2) \bmod p \neq 0 \bmod p \qquad (2)$$

The elliptic curve E(a, b) over the prime field Fp consists of set of points (x, y) which satisfy Eqn. (1) and it has one additional element called point at infinity **O**. E(a, b) is a group, which is closed under addition operation:

1. Point at infinity **O** is the additive identity, P+**O** = **O**+P = P.
2. The negative of P(x, y) is –P(x,–y), i.e., the same x-coordinate with negative of y-coordinate. These two points P and –P can be joined by a vertical line and meet at point at infinity (Figure 8.1), i.e.:

$$P + (-P) = P - P = \mathbf{O}$$

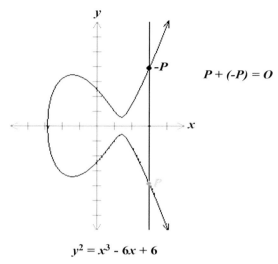

$$y^2 = x^3 - 6x + 6$$

FIGURE 8.1   Point Addition P + (–P) = O.

3. Two different points P and Q can be added, P + Q = –R, which means P+Q is the point of the mirror image of the third point of intersection R (Figure 8.2).

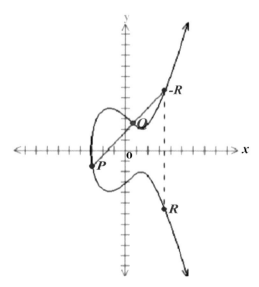

FIGURE 8.2   Point addition P + Q = –R.

4. To double a point Q, a tangent line is drawn and the other point of intersection is S then P + P = 2P = –R (Figure 8.3).

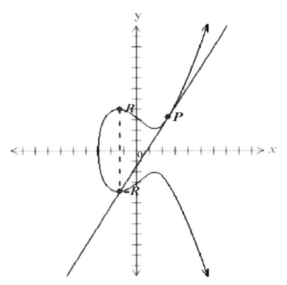

FIGURE 8.3 Point doubling 2P = –R.

If P and Q are two points of the curve with coordinates (x_1, y_1) and (x_2, y_2) with P ≠ –Q, then the point R=P+Q with coordinates (x_3, y_3) is computed as:

$$x_3 = (\lambda^2 - x_1 - x_2) \, modp \qquad (3)$$

$$y_3 = (\lambda \, (x_1 - x_3) - y_1) \, modp \qquad (4)$$

where, λ calculated with the formula:

$$\lambda = \begin{cases} \left(\dfrac{y2 - y1}{x2 - x1} \right) modpif \ P \neq Q \\[3mm] \left(\dfrac{3x1^2 + a}{2y1} \right) modpif \ P = Q \end{cases} \qquad (5)$$

The basic operations on elliptic curve are point addition and point doubling. The scalar point multiplication operation that means if k is a positive integer and P is a point on the elliptic curve, then kP = P + P + P.... (k times) required to be calculated. This can be done with a sequence of point addition operation and point doubling operation.

The number of points (x, y) which fulfill the equation of elliptic curve given by Eqn. (1) along with the point at infinity, O is called the order of the elliptic curve #E. Order of a point G on the elliptic curve is defined as a value n such that:

$$nP = P + P + P + \ldots \text{(n times)} = O \text{ (point at infinity)}$$

8.3.2 ELLIPTIC CURVE PUBLIC KEY CRYPTOSYSTEM

In the public key cryptosystem each participant in the communication has two keys; private key which has to be kept as a secret, it is also called secret key, the other key is called public key which can be shared among other participants of communication. Both the sender and the receiver must mutually agree upon the domain parameters (a, b, p, G, n). Where a and b are the parameters of Elliptic Curve equation E(Fp), p is the prime number, G is the generator point on the curve and n is the order of G.

The sender generates a random number $n_A < n$, then sender has to calculate his/her public key $P_A = n_A.G$, where the sender's private key is n_A.

The receiver generates a random number $n_B < n$, then the receiver has to calculate his/her public key $P_B = n_B.G$, n_B is the private key of the receiver.

1. **ElGamal Encryption:** The message m which needs to be encrypted has to be encoded as an elliptic curve point, let it be P_m. The sender has to select a random number k < n. The cipher text Cm generated is a point pair $\{kG, P_m+kP_B\}$ where P_B is the public key of the receiver.

2. **ElGamal Decryption:** The Ciphered text is a pair $\{kG, P_m+kP_B\}$. The receiver multiplies the private key n_B with the first point of the point pair and subtracts the result from the second point.

$$\text{i.e., } (P_m + kP_B) - n_B.kG = P_m + kP_B - kP_B = P_m$$

The Pm can be decoded to the original message m.

8.3.3 ARNOLD'S TRANSFORMATION

Vladimir I Arnold explained Arnold transformation, popularly called cat map. It is applicable over various data types in spatial domain; hence it is used on other domains like scrambling of image and image encryption.

$$\begin{bmatrix} x' \\ y' \end{bmatrix} = \begin{bmatrix} 1 & 1 \\ 1 & 2 \end{bmatrix} \begin{bmatrix} x \\ y \end{bmatrix} \text{mod n} \tag{6}$$

where, x and y are the original values and x' and y' are the new scrambled values. The generalized Arnold transformation is also called modified Arnold transformation:

$$\begin{bmatrix} x' \\ y' \end{bmatrix} = \begin{bmatrix} 1 & a \\ b & ab+1 \end{bmatrix} \begin{bmatrix} x \\ y \end{bmatrix} \text{mod n} \tag{7}$$

where, a, b, and n are considered as the keys of Arnold transformation. The reverse transformation of Arnold transformation is given as:

$$\begin{bmatrix} x \\ y \end{bmatrix} = \begin{bmatrix} ab+1 & -a \\ -b & 1 \end{bmatrix} \begin{bmatrix} x' \\ y' \end{bmatrix} \text{mod n} \tag{8}$$

8.4 THE PROPOSED SCHEME

The selected elliptic curve equation is $y^2 \bmod 239 = (x^3 + 2x + 1) \bmod 239$, along with the point at infinity this EC has 257 points. The generator point selected as G = (1, 2) whose order is 257 as 257 is a prime number.

An image is represented by a matrix of integers. In a grayscale image, the value of each pixel value can be expressed by 8 bits. The 8-bit can represent 256 grayscale values (integer values ranging from 0–255). We first apply Arnold's transformation to produce confusion and diffusion. We have selected the values of a = 3 and b = 7 in Eqn. (7).

Then each value of the scrambled pixel intensity is encrypted by using ElGamal method on elliptic curve cryptography. We have selected an elliptic curve equation such that it could generate 257 points including point at infinity. All generated points of the elliptic curve are stored in an array list of Java, whose index starts with 0, we encode the transformed pixel intensity with the point in the array list which has the corresponding index value.

After getting Pm the ElGamal encryption on EC is applied and it produced the ciphered text Cm which is a pair of points. Then the ciphered image is sent through the network to the receiver.

The receiver applied ElGamal decryption on EC to get the transformed pixel intensity. Then the Arnold's reverse transformation is applied to get the original image by using Eqn. (8) (Figures 8.4 and 8.5).

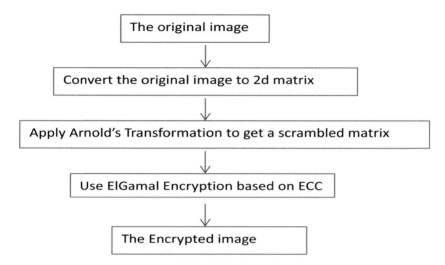

FIGURE 8.4   Sender side operations.

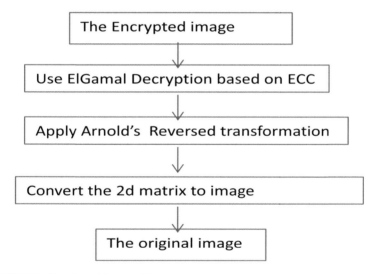

FIGURE 8.5   Receiver side operation.

8.5   SIMULATION RESULT

We have chosen some grayscale images, image of baboon size 225 x 225 and image of Lena size 512 x 512 in .bmp format, image of cameraman

.jpg format size 225 x 225, boat image in .png format size 512 x 512 for our experiment. All the experiments were conducted by using Java and MATLAB R2015 a software was used to analyze our proposed scheme (Figures 8.6–8.9).

FIGURE 8.6 Sample 1: (a) Image before encryption; (b) image after encryption; (c) image after decryption.

FIGURE 8.7 Sample 2: (a) Image before encryption; (b) image after encryption; (c) image after decryption.

FIGURE 8.8 Sample 3: (a) Image before encryption; (b) image after encryption; (c) image after decryption.

FIGURE 8.9 Sample 4: (a) Image before encryption; (b) image after encryption; (c) image after decryption.

The points generated on $y^2 \bmod 239 = (x^3 + 2x + 1) \bmod 239$ (Table 8.1).

TABLE 8.1 Points on Elliptic Curve $y^2 \bmod 239 = (x^3 + 2x + 1) \bmod 239$

Index	Point
0	(0, 1)
1	(0, 238)
2	(1, 2)
3	(1, 237)
4	(3, 89)
5	(3, 150)
–	–
–	–
253	(232, 220)
254	(233, 27)
255	(233, 212)

8.6 SECURITY ANALYSIS

1. **Key Space Analysis:** Elliptic curve cryptography is recognized for the capability to provide high security. Compared to other crypto-systems the elliptic curve discrete logarithm problem (ECDLP) provides better security even for small key size. In our

proposed scheme key space for modified Arnold transformation is K1 and key space for Elliptic curve cryptography is K2. Then the total key space for our scheme is K1 x K2, where K1 = 256 x 256 and K2 = 256!

2. **Key Sensitivity Analysis:** An algorithm is called secured if it is completely sensitive to the secret key. With a minor change in the secret key the encrypted image should not be decrypted. Changing even a single bit in secret key would not produce the correct result. Figure 8.10 reveals that by changing a single bit of secret key we are not getting the correct result, hence our scheme is resistant to brute force attack.

FIGURE 8.10 (a) Original image; (b) encrypted image; (c) decrypted with correct key; (d) decrypted with wrong key.

3. **Information Concealing:** If encryption process is done by the receiver's public key and decryption process is done by the receiver's private key then the image can be recovered lossless. Figure 8.6(a–c) depicts the original image, the encrypted image, and the decrypted image of Lena. Figure 8.7(a–c) depicts the original image, the encrypted image, and the decrypted image of Baboon. Figure 8.8(a–c) depicts the original image, the encrypted image, and the decrypted image of Cameraman. Figure 8.9(a–c) shows the original image, the encrypted image, and the decrypted image of Boat. It is clearly visible that the information of all the plain images is entirely concealed.

4. **Known Plain Text Attack:** If the intruder knows the ciphered text, encryption algorithm and some plaintext-ciphered text pairs produced using a secret key, may be able to get the secret key. The idea of using ECC for encryption produces a completely different cipher text using the same key in every execution, because cipher

text is dependent on the value of k which is a randomly generated value.

5. **Statistical Analysis:** Three types of statistical analysis were performed:

    i.   **Histogram Analysis:** The histogram of gray levels' of a digital image is within the range [0, P-1] which is a discrete function $f(m_k) = p_k$, the kth gray level of the image is m_k and the count of pixels in the image that have m_k gray level is p_k. Confusion is meant for making the association between the key and cipher text as complicated as possible. The adversary can't figure out the key for encryption from cipher text. In our simulation, the histograms of the plain images and cipher images are computed and shown in Figure 8.11. In the graphs, the pixel values are represented in x-axis the number of pixels is represented in y-axis. Figure 8.11(a) is the histograms of the grayscale Lena image. Figure 8.11(b) is the histogram of the image of Lena after encryption. Figure 8.11(c) is the histogram of the image of Baboon, whereas Figure 8.11(d) is the histogram of the image of baboon after encryption. Figure 8.11(e) is the histogram of the image of original boat image before encryption whereas Figure 8.11(f) is the histogram of the image of boat after encryption.

        The histogram of the ciphered images describes the uniform distribution characteristics whereas the histograms of the plain images do not exhibit this characteristic. The histograms of the encrypted images are remarkably different from the histograms of the original images. This result shows the confusion property of our scheme.

    ii. **Correlation Coefficient Analysis:** We have conducted a correlation coefficient analysis by randomly choosing 3000 adjacent pixel pairs in all three directions: Horizontal direction, vertical direction, and diagonal direction from the original or plain image of Lena and the encrypted or ciphered image of Lena. The coefficients of correlation of both the original image and encrypted images in horizontal, vertical, and diagonal directions are revealed in Figure 8.12(a–f) and Table 8.2.

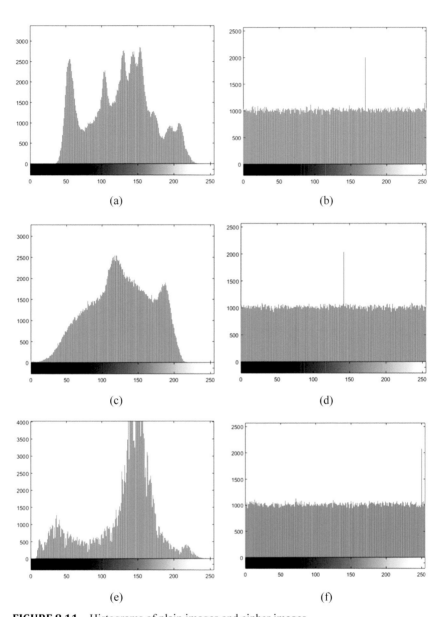

FIGURE 8.11   Histograms of plain images and cipher images.

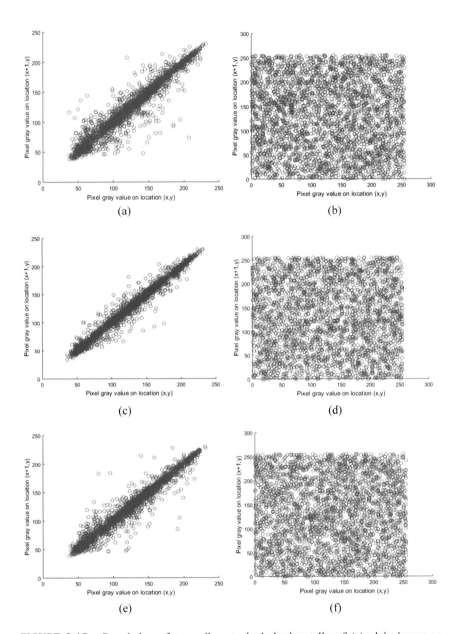

FIGURE 8.12 Correlation of two adjacent pixels horizontally of (a) plain image or original image (b) ciphered image. Correlation of two adjacent pixels vertically (c) plain image or original image (d) ciphered image. Correlation of two adjacent pixels diagonally of (e) plain image or original image (f) ciphered image.

TABLE 8.2   Correlation Coefficients of 2 Adjacent Pixels in the Original Image and Encrypted Image

	Horizontal	Vertical	Diagonal
Lena image (original)	0.9625	0.9878	0.9745
Lena image (encrypted)	–0.0217	0.0034	–0.0198
Baboon image (original)	0.8599	0.8010	0.7403
Baboon image (encrypted)	–0.0140	0.0038	–0.0238
Boat image (original)	0.9370	0.9745	0.9267
Boat image (encrypted)	0.0265	0.0344	–0.0355
Cameraman image (original)	0.9435	0.9648	0.9161
Cameraman image (encrypted)	0.0120	0.0189	0.0159

$$r_{xy} = \frac{\text{cov}(x, y)}{\sqrt{D(x)}\sqrt{D(y)}},$$

where,

$$\text{cov}(x, y) = \frac{1}{N}\sum_{i=1}^{N}(x_i - E(x))(y_i - E(y)),$$

$$D(x) = \frac{1}{N}\sum_{i+1}^{N}(x_i - E(x))^2, \; E(x) = \frac{1}{N}\sum_{i=1}^{N}x_i.$$

where, x and y represent the gray-scale values of two adjacent pixels.

We have also analyzed the correlation between the images before and after encryption. The coefficient correlation between original and encrypted images is close to 1, i.e., both the images are same without any alteration of the value of the pixels.

iii. **Information Entropy:** Entropy is a fine measure of randomness. The encrypted image is not predictable or random if the value of entropy is calculated to be around 8 and hence it is said to provide high level of security. All our ciphered images provide entropy around 8. Hence we can say that our scheme provides sufficient randomness. Shannon's entropy function is used to calculate the entropy value. To calculate entropy, the given formulae are used (Table 8.3).

$$E_h = \sum_{j=1}^{l^n} p(nj)\log_2 p(nj)$$

TABLE 8.3   Entropy Values

Images	Original	Ciphered
Lena	7.3479	7.9972
Baboon	7.4050	7.9972
Boat	7.1914	7.9969
Cameraman	7.0629	7.9942

The probability of n_j pixel sequence among 256 values is represented as. Our scheme used the given values: $l = 2$, $n = 8$ (binary of each pixel value).

8.7   CONCLUSION AND FUTURE WORK

In our chapter, we presented the scheme for encryption based on Arnold's transformation and the ElGamal encryption method based on Elliptic curve cryptography. We worked on grayscale images. Our proposed scheme is based on changing the values of the pixels a given image by using Arnold transformation; then using the difficulty of ECDLP we encrypt the scrambled image using the ElGamal method of ECC. The scheme is a lossless scheme too. The proposed scheme may be extended for color images. In our future work, we may use the digital signature of elliptic curve cryptography to ensure authentication.

KEYWORDS

- **Arnold's cat map**
- **decryption**
- **discrete logarithm problem**
- **ElGamal algorithm**
- **elliptic curve**
- **encryption**
- **image encryption**
- **private key**
- **public key**

REFERENCES

1. El-Latif, A. A., & Xiamu, N., (2013). A hybrid chaotic system and cyclic elliptic curve for image encryption. In: *AEU-International Journal of Electronics and Communications* (Vol. 67, No. 2, pp. 136–143). Elsevier.
2. Ali, S., Jan, N. M., & Zulkarnain, M. A., (2013). A Novel public key encryption based on elliptic curves over prime group field. *Journal of Image and Graphics*, 1, 43–49.
3. Ann, H. K., Neal, K., & Alfred, M., (2011). Elliptic curve cryptography: The serpentine course of a paradigm shift. *Journal of Number Theory*, *131*, 781–814. Elsevier.
4. Binay, K. S., Abusha, T., & Jagat, S., (2017). *Probabilistic Data Encryption Using Elliptic Curve Cryptography and Arnold Transformation*. I-SMAC.
5. Darrel, H., Alfred, M., & Scott, V., (2004). *Guide to Elliptic Curve Cryptography*. Springer.
6. Don, J., Alfred, M., & Scott, V., (2001). *The Elliptic Curve Digital Signature Algorithm (ECDSA)*. Certicom Corporation.
7. Hong, L., & Yanbing, L., (2014). Cryptanalyzing an image encryption scheme based on hybrid chaotic system and cyclic elliptic curve. In: *Optics and Laser Technology* (Vol. 56, pp. 15–19). Elsevier.
8. Laiphrakpam, D. S., & Khumanthem, M. S., (2015). Image Encryption using elliptic curve cryptography, eleventh international multi-conference on information processing–2015 (IMCIP-2015). *Procedia Computer Science, 54*, 472–481.
9. Lawrence, C. W., (2008). *Elliptic Curves Number Theory and Cryptography* (2nd edn.). Taylor & Francis Group.
10. Li, L., El-Latif, A. A. A., & Xiamu, N., (2012). Elliptic curve ElGamal based homomorphic image encryption scheme for sharing secret images. In: *Signal Processing* (Vol. 92, pp. 1069–1078). Elsevier.
11. Neal, K., (1987). Elliptic curve cryptosystems. *Mathematics of Computation*, *48*(177), 203–209.
12. Neal, K., Alfred, M., & Scott, V., (2000). The state of elliptic curve cryptography. *Designs, Codes and Cryptography, 19*(2/3), 173–193.
13. Omar, R., (2018). Text message encoding based on elliptic curve cryptography and a mapping methodology. *Inf. Sci. Lett., 7*(1), 7–11.
14. Behnia, S., Akhavan, A., Akhshani, A., & Samsudin, A., (2013). Image encryption based on the Jacobian elliptic maps. In: *The Journal of System and Software* (Vol. 86, pp. 2429–2438). Elsevier.
15. Maria, S., Celestin, V., & Muneeswaran, K., (2012). Nonce based elliptic curve cryptosystem for text and image applications, In: *International Journal of Network Security* (Vol. 14, No. 4, pp. 236–242).
16. Victor, S. M., (2000). Use of elliptic curves in cryptography. *Advances in Cryptology-RYPTO '85 Proceedings* (Vol. 218, pp. 417–426). Springer.
17. Williams, S., (2000). *Cryptography and Network Security* (4th edn.). Prentice Hall, Pearson.

CHAPTER 9

Security Management in SDN Using Fog Computing: A Survey

MANISH SNEHI and ABHINAV BHANDARI

Department of Computer Engineering, UCOE, Punjabi University, Patiala, Punjab, India

ABSTRACT

Traditional networks, despite of its popularity adopted network schemas which are difficult to maintain and manage. Software-defined networking (SDN) is an emerging trend which separates the control plane and data plane in the network. SDN has gained attention of researchers due to its ability to easily scale, extend, and dynamically add network functions. The dynamic capabilities of SDN have brought about a significant change in the way networks were implemented. To make SDN-based networks more reliable and stable, significant effort from researchers' is required. In this chapter, the attention is given to security concerns in SDN networks, especially DDoS attacks. Distributed denial of service (DDoS) attack is one of the serious security attacks which make a resource unavailable to its legitimate users by flooding or consuming entire network bandwidth. The survey highlights Fog computing as an additional layer for reducing latency between the perception layer and cloud as well as to address the security concerns such as detecting DDoS attacks. Fog computing layer leverages SDN controller to gather traffic data, detect DDoS attacks, and mitigate the same. The review is helpful in understanding how SDN can be used in Fog Computing layer to suppress DDoS attacks.

9.1 INTRODUCTION

Tight coupling of vertical planes in traditional network devices (control plane and data plane) makes it difficult to implement predefined network policies. The difficulty is caused because the configuration of routers and switches is not flexible enough to address the frequently changing needs to respond to traffic loads and fault tolerance in the network [1]. Despite of high adoption of traditional network components, the network topology is complex and hard to manage [1, 2]. In addition, the configuration of network components from multiple vendors requires low-level vendor-specific commands to configure switches and/or routers, which demands highly skilled professionals to manage the network infrastructure.

Software-defined networking (a.k.a. SDN) is an emerging technology trend which addresses the issues and challenges in traditional networking system [3, 4] by focusing on:

1. Decoupling of vertical integration of data plane and control plane.
2. The high-level view of the network and centralized control of devices is visible at the point of control.
3. Standardized protocols have been devised for control plane and data plane communication.
4. Making it highly extensible and configurable because of supporting middle boxes in the network infrastructure.

The control and data plane separation are realized by programming interfaces between the SDN controller and router/switches. As the SDN controller is built on top of the network operating system, it is highly extensible to support a number of middle box configurations like intrusion detection systems, WAN Optimizers, Load Balancers, or network address translators (NAT). The middle boxes are either part of SDN controller or have their existence on the virtual environment, and SDN controller leverages the benefits provided by middle boxes by means of network function virtualization (a.k.a. NFV). Figure 9.1 [5] presents software-defined networking architecture illustrating application, control, data layers, and corresponding interfaces.

However, benefits offered by SDN come with a cost of security challenges associated with software-defined networks. One of the serious kinds of security attacks are distributed denial of service (DDoS) attacks, which makes the application or resource unavailable to the legitimate user.

For example, some of the devices are used to flood the network bandwidth by generating massive traffic data so that the application is unavailable to genuine users. The centralized controller and flow table limitations act as weak entry gates for DDoS attacks [5]. In addition to DDoS attacks, other security concerns faced by SDN configuration are based on compromising the trusted relationship between SDN controller and application hosted on the SDN controller due to open programmability of network.

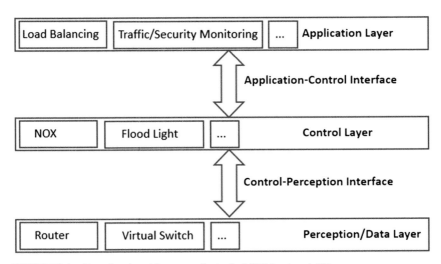

FIGURE 9.1 Functional architecture of a typical SDN network [5].

In modern SDN-enabled application architecture where the data received by switches or routers is passed on either to the cloud or any other network enabled application, the early detection of DDoS attacks is of utmost importance.

Fog assisted SDN is additional layer in between perception layer where the data producer devices like sensors and/or IoT devices are configured and cloud computing layer where the data from devices is destined [6]. The fog layer is comprised of SDN infrastructure as one of its primary components.

The primary contribution of survey is described as under:

a. We present the logical and organized review of SDN architecture and analyze various security challenges in software-defined networking.
b. In this survey, we present the role of fog computing in SDN to overcome the security challenges.

The rest of the chapter is organized as follows: Section 9.2 presents literature survey; Section 9.3 describes open challenges and issues in tradition networking system; and Section 9.4 explains how fog computing is used to leverage SDN architecture and addresses the security concerns. The last Section 9.5 concludes the survey.

9.2 LITERATURE SURVEY

SDN due to its tremendous growth is attracting the attention of both industry and academia. The Open Networking Foundation (ONF) [7] is dedicated to develop and standardize SDN [8]. OpenFlow protocol is an important element of SDN-based networks.

SDN can improve the IoT ecosystem by providing better manageability, controllability, and scalability of modern SDN-based networks. The authors of [9] infer that the network model, which provides connectivity and abstraction on applied network policies, are the backbone of the Cloud computing layer. They introduce Meridian, an SDN-based networking framework, which supports the application-networking model at the service level and is used to implement various options of virtual networks.

Yen et al. [10] established a Cloud-based computing environment with SDN controller and OpenFlow protocol running in its core network components. Further, the functionality was extended to Load Balancers and various traffic monitoring applications.

Ku et al. [11] showcased SDN-based mobile cloud schemas. Their focus area was ad hoc networks. Gharakheili et al. [12] proposed architecture comprised of SDN interfaces in the back-end logic and front-end is taking advantage of Cloud platform. The architecture is used by ISPs to allow users to customize the network.

Although SDN-based networks offer a variety of features in the list, there are challenges which must be addressed including performance issues, ensuring network availability, scalabilities issues, and security implementation.

Jarschel et al. [27] demonstrated that big number of flows can't be handled by current controllers in 10 Gbps links. Hence, the performance was one of the key areas which required further research attention.

As discussed, SDN-based networks require special attention to security implementation because of the following (but not limited to) threats [13, 14]:

1. Forged traffic flows;
2. Attacks on data plane components;
3. Attacks on control plane (SDN controller);
4. Attacks due to lack of ways to ensure a trusted relationship between hosted application on controller and controller.

A number of defense schemes are presented in the literature [15–17], Farahmandian et al. focused on defense mechanisms against DDoS attacks in Cloud environment [16].

Yan et al. [6] proposed SDN-based implementation in Fog-Computing layer to detect and mitigate the DDoS attacks in timely manner, well before the malicious traffic flows to the Cloud layer of the IoT eco-system.

9.3 OPEN RESEARCH ISSUES IN SOFTWARE-DEFINED NETWORKING

SDN provides the capability to program and dynamically manage the network infrastructure. The network policies can be created at run-time on the controller and some are enforced on to network switches and/or routers [5].

As SDN is an emerging technology trend which exposes programmable interfaces to network infrastructure; a number of researchers are focusing on preventing the security attacks on the SDN controller itself. In addition, a lot of effort is made in research community to use SDN as a solution or tool to enhance the security of SDN-enabled applications [6, 8, 18].

There are a number of open issues related to network security, specifically DDoS attacks, which have got researchers' attention. DDoS attacks results in non-availability of an application or network components based on the targeted area for DDoS attacks.

As SDN itself is prone to attacks because of its architecture [1], potential attackers can target any or all the three layers to make SDN a victim of the attacks. Hence, DDoS attacks on SDN can be classified into: Application layer attacks, Control layer attacks, and Data plane layer attacks. Figure 9.2 [5] describes the various layers in SDN architecture which could be the potential target for compromising security of a layered network system.

In each of the category of attacks, either the specific components residing on the corresponding layer is targeted or the interface which is extended by SDN for communicating to another layer (upward or downwards) is targeted. For example, in Application layer DDoS attacks, an application or a set of applications on SDN controller are beset so as to result in SDN controller unavailability (DOS). The attack focus area can be the Communication Interface such as Southbound APIs, Northbound APIs, Eastbound, or Westbound APIs [19].

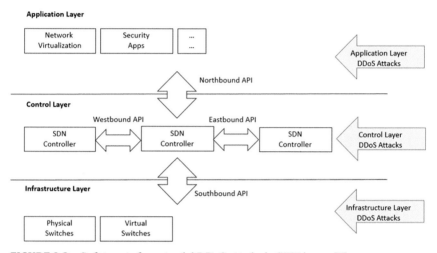

FIGURE 9.2   Soft targets for potential DDoS attacks in SDN layers [5].

There are researches in each of the direction specific to compromised network component. In addition, there are open issues which are waiting to be addressed. In this section, we further discuss some of the important research issues to mitigate the DDoS attacks in cloud environment of SDN:

1. **Defeating Application Layer Attacks Using SDN:** As mentioned earlier, application layer DDoS attacks are quite smart category of attacks because they consume limited bandwidth and focuses on taking an advantage of request volumes for a specific application [11] thereby making the application unavailable for legitimate users.

    Gartner, in his research, stated that there is a significant growth in application-level DDoS attacks [20]. Major efforts are required to be spent in order to provide effective solutions which should take care of trade-offs between security aspects of SDN and network performance.

2. **Addressing Mobile Security Attacks using SDN:** A survey done by Prolexic Technologies reported that the mobile applications

are a major source of DDoS attacks [21]. An effective mitigation framework requires additional measures to be taken to control the attacks. As an example, inclusion of fingerprinting, captcha images and other such unique methods which are hard to replicate by attackers can be a great help [22].

3. **Building DDoS Tolerant Systems by Leveraging SDN Functionality:** The work carried out for mitigating DDoS attacks conveys the difficulty in accurately detecting DDoS attacks and most importantly, stopping them in early phase of attack [23]. Therefore, the designing of such a fault tolerant system is required which can operate efficiently and correctly to withstand against DDoS attacks.

4. **Leveraging Fog Computing as Container for SDN Infrastructure:** Fog computing is an additional layer introduced between the data perception layer (where the data is generated) and the data sink layer (where the data is destined; for example, cloud) [24].

As compared to cloud, Fog possesses following characteristics:

a. Major amount of data storage and analytics to be carried our near end users;

b. Deals with majority of communication and networking near end users.

Various benefits of Fog are summarized as: Cognition, agility, and addressing the latency.

9.4 FOG COMPUTING IN SDN TO ADDRESS SECURITY CONCERNS

Fog can support the cyber-physical systems (CPS) and an addition of SDN schema into the fog network can allow monitoring; and analytics can be applied to the data traffic before data reaches to its destination (i.e., Cloud) [6].

With the evolution of NFV, fog layer can be configured to install middle boxes in terms of NFV functions to support dynamic functionality to SDN enabled network. Figure 9.3 [6] shows the high-level architecture of an IoT eco-system using SDN enabled cloud to address security concerns.

The research work is carried out in the direction of enabling the SDN controller to handle a number of security concerns by collecting the traffic data, applying analytics and machine learning algorithms to detect DDoS attacks, suppressing the DDoS attacks, and then sending the clean data from legitimate end users to the cloud.

Security functions can be embedded into SDN controller's application because of its ability to provide programmable interfaces and capability to host network applications [16, 25].

9.5 CONCLUSION

There are two branches of research thoughts for discussing about the security in Software-defined networking. The first branch focuses on network security achieved by exploring the various methods to exploit the programmability and centralized control offered by SDN. The second thought engages the SDN as a tool to address the security concerns and introduce Fog computing layer between data producer devices and data consumer infrastructure.

Our survey identifies that Fog computing layer to contain SDN infra-structure is one of the areas where research is able to catch the attention of researchers. A concerted effort in both the areas, securing SDN as well as using SDN to secure network, could yield a more trusted and reliable SDN.

ACKNOWLEDGMENTS

The authors would like to thank to the anonymous reviewers for their feedback and suggestions.

KEYWORDS

- **distributed denial of service**
- **fog computing**
- **network address translators**
- **network function virtualization**
- **programmable networks**
- **software-defined networking**

REFERENCES

1. Kreutz, D., Ramos, F. M. V., Veríssimo, P. E., Rothenberg, C. E., Azodolmolky, S., & Uhlig, S., (2015). Software-defined networking: A comprehensive survey. *Proc IEEE., 103*(1), 14–76.

2. Sezer, S., et al., (2013). Are we ready for SDN? Implementation challenges for software-defined networks. *IEEE Commun. Mag., 51*(7), 36–43.

3. Mckeown, N., (2011). *How SDN Will Shape Networking.*

4. Schenker, S., (2011). *The Future of Networking the Past of Protocols.*

5. Scott-Hayward, S., O'Callaghan, G., & Sezer, S., (2013). "SDN security: A survey. *Proc. IEEE SDN Future Netw. Services* (pp. 1–7).

6. Yan, Q., et al., (2018). A multi-level DDoS mitigation framework for the industrial internet of things. *IEEE Commun. Mag., 56*(2), 30–36.

7. Open Networking Foundation, (2014). [Online] Available at: https://www.opennet-working.org (accessed on 3 November 2020).

8. Yan, Q., Yu, F. R., Gong, Q., & Li, J., (2016). Software-defined networking (SDN) and distributed denial of service (DDoS) attacks in cloud computing environments: A survey some research issues and challenges. *IEEE Commun. Surv. Tutor., 18*, 602–622.

9. Banikazemi, M., Olshefski, D., Shaikh, A., Tracey, J., & Wang, G., (2013). Meridian: An SDN platform for cloud network services. *IEEE Commun. Mag., 51*(2), 120–127.

10. Yen, T. C., & Su, C. S., (2014). An SDN-based cloud computing architecture and its mathematical model. *Proc. IEEE Int. Conf. ISEEE, 3*, 1728–1731.

11. Ku, I., Lu, Y., & Gerla, M., (2014). Software-defined mobile cloud: Architecture services and use cases. *Proc. IEEE IWCMC*, 1–6.

12. Gharakheili, H., Bass, J., Exton, L., & Sivaraman, V., (2014). Personalizing the home network experience using cloud-based SDN. *Proc. IEEE Int. Symp. WoWMoM*, 1–6.

13. Bhandari, A., Sangal, A. L., & Kumar, K., (2016). Characterizing flash events and distributed denial-of-service attacks: An empirical investigation. *Security and Communication Networks.*

14. Bhandari, A., Sangal, A., & Kumar, K., (2014). Performance metrics for defense framework against distributed denial of service attacks. *International Journal on Network Security, 5*(2), 38.

15. McKeown, N. (2012). *RioRey Taxonomy of DDoS Attacks* [online] Available at: http://www.riorey.com/x-resources/2012/RioReyTaxonomyDDoSAttacks2012.eps (accessed on 3 November 2020).

16. Farahmandian, S., et al., (2013). A survey on methods to defend against DDoS attack in cloud computing. *Proc. Recent Adv. Knowl. Eng. Syst. Sci.*, 185–190.

17. Zargar, S. T., Joshi, J., & Tipper, D., (2013). A survey of defense mechanisms against distributed denial of service (DDoS) flooding attacks. *IEEE Commun. Surveys Tuts., 15*(4), 2046–2069.

18. Yan, Q., et al., (2016). Software-defined networking (SDN) and distributed denial of service (DDoS) attacks in cloud computing environments: A survey some research issues and challenges. *IEEE Commun. Surveys and Tutorials, 18*(1), 602–22.

19. Sezer, S., et al., (2013). Are we ready for SDN? Implementation challenges for software-defined networks. *IEEE Commun. Mag., 51*(7), 36–43.

20. Shenker, S., Martin Casado, Teemu Koponen, & Nick McKeown (2013). *Gartner: Application Layer DDoS Attacks to Increase in 2013* [online].

21. Yan, Q., Wenyao Huang, Xupeng Luo, Qingxiang Gong, & Richard Yu, F. (2014). *Mobile Applications being Used for DDoS Attacks According to Prolexic's Latest Quarterly Report* [online] Available at: http://www.prolexic.com/news-events-pr-mobile-apps-applications-being-used-for-ddos-attacks-q4-2013-ddos-attack-report.html (accessed on 3 November 2020).

22. Sun, B., Yu, F., Wu, K., Xiao, Y., & Leung, V., (2006). Enhancing security using mobility-based anomaly detection in cellular mobile networks. *IEEE Trans. Veh. Technol., 55*(4), 1385–1396.

23. Gashi, I., & Kreidl, O. P., (2012). 6th workshop on recent advances in intrusion tolerance and resilience (WRAITS 2012). *Proc. 42nd IEEE/IFIP Int. Conf. DSN-W*, 1, 2.

24. Chiang, M., & Zhang, T., (2016). Fog and IoT: An overview of research opportunities. *IEEE Internet Things J., 3*(6), 854–864.

25. Yan, Q., & Yu, F. R., (2015). Distributed denial of service attacks in software-defined networking with cloud computing. *IEEE Commun. Mag., 53*(4), 52–59.

26. Farahmandian, S., et al., (2013). A survey on methods to defend against DDoS attack in cloud computing. *Proc. Recent Adv. Knowl. Eng. Syst. Sci.,* 185–190.

27. Jarschel, M., Oechsner, S., Schlosser, D., Pries, R., Goll, S., & Tran-Gia, P. (2011). "Modeling and performance evaluation of an OpenFlow architecture," *23rd International Teletraffic Congress (ITC)*, San Francisco, CA, pp. 1–7.

CHAPTER 10

Comparative Evaluation of Free/Open Source and Proprietary Software Using Survey of Indian Users

SUSHIL KUMAR and SUMESH SOOD

Department of Research, Innovation and Consultancy, I. K. Gujral P.T.U, Kapurthala, Punjab, India

ABSTRACT

Although much research work has been done on free and open-source software at the international level but in context of Indian users much work is not done yet. Measuring the satisfaction level of theses software users is a key issue which must be explored on the basis of users' response to this software. The same has been done in the present chapter. In addition, the collected data has been processed using SPSS to find out to which category of software, i.e., FOSS (free/open source software) and proprietary the users' response is positive and why. To some factors response was found to be positive but in other factors it was found to be negative. The present chapter explores the overall conclusion from data processing.

10.1 INTRODUCTION

The objective of this chapter is to compare proprietary and free/open-source software by applying descriptive frequency test and one sample T-test using SPSS on the data collected through questionnaire survey from Indian users. The prime motive of the present research work is to investigate responses of FOSS users and help the government in policy formulation on adoption of FOSS software. Even many payment schemes that currently exist are proprietary, insecure, inflexible or all three [1].

FOSS may be better alternative in such a scenario. This research chapter incorporates variety of aspects of technology like human interaction with computers using proprietary software versus FOSS, technology assessment of software development, i.e., proprietary software versus FOSS, social impact of software development using proprietary software versus FOSS development methodologies, technology enhancement learning using proprietary software versus FOSS and social shaping of technology using two categories of software, i.e., proprietary software and FOSS. The questionnaires [2] in form of Google forms for users at distance and printed questionnaire has played a good role to collect the users' response from local users on this software. SPSS [3] has been used for database table creation and entering into it, users' response to the questionnaires. The source of questions is given in Ref. [4]. The overall results have been obtained by processing these responses with descriptive frequency tests and one sample T-test using SPSS (statistical package for social sciences). As there is no experimental set up without microprocessor control [5], all the collected data has necessarily to be processed on computers.

10.1.1 STATISTICAL PACKAGE FOR THE SOCIAL SCIENCE (SPSS)

Statistical package for the social science (SPSS) software runs on Windows [6]. SPSS takes data from files created in various types of software packages. It uses them to generate tabulated reports, charts, distributions, and trends, descriptive statistics, and complex statistical analysis. SPSS is capable of handling large amounts of data and can perform versatile type of analysis. SPSS is used in the business world and analysis work in research problems of various types of disciplines. SPSS is a popular statistical package. It performs data manipulation and analysis with simple instructions.

10.2 METHODOLOGY AND SOURCES OF DATA

Steps of algorithm covered for completing this research model's processes have been conducted throughout the world but it has not happened in context of India before. And to select the standard questions on basis of standard factors for comparison between two types of software the set of questions have been taken from Framework for OSS adoption, Government of India, April 2015.

1. In this step Google form named "Questionnaire for Comparison of Proprietary Software with FOSS in Context of India from the viewpoint of Indian Users" was developed. In addition, the same questionnaire was also taken in printed form to distribute among users of both types of software.
2. This survey was conducted from November 2015 to April 2018 from some states of India. E-mails were sent to 1000 probable participants in many organizations. To increase the survey response rate, a paper-based questionnaire was also sent to 1000. Overall nearly 2000 invitations were sent for this survey. Responses from more than 800 participants were received. This information was collected from individuals involved in software organizations and educational organizations. The response rate was good.
3. The respondents of the online Google form of questionnaire were selected from the mailing lists of PhD Scholars, e-mail Ids of participants of computer conferences, Professors of Computers, school teachers of computer, students mostly of computer branches. SPIC India (The Society for Promotion of IT in Chandigarh) Department from IT Park, Chandigarh provided list of teachers who has experience of Bharat Operating System Solution in computer laboratories of schools provided by C DAC Mohali. The author traveled so many computer institutes to get the printed questionnaires filled by computer staff and the students.
4. The author has also joined the WhatsApp group of BOSS users of all over India named "Bharat operating system" and got filled some online questionnaires from them.
5. The author is also a member of https://groups.google.com/ forum/#!forum/boss-support here one can easily consult regarding any query regarding Bharat Operating System Solution.
6. Developing the questionnaire for survey is a tedious process. The questionnaire for this survey was developed using the factors mentioned in Framework for FOSS adoption by IT Department of Government of India.
7. Question types used in the questionnaire: The most of the questions in questionnaire are closed-ended. Closed-ended questions are effectively designed with comprehensive answer choices. This type of questioning is very useful in collecting information to

support theories and concepts in the literature. Closed-ended questions are constructed using multiple choices intend to obtain a single response for each question. The choice criterion is clearly defined to cover the maximum possible answers for the questions in the questionnaire.

8. Construction of questionnaire, the questionnaire is developed based on the factors reviewed from the literature, the questions comprise cost of software, user friendliness of software, freedom to use Software, piracy of Software, data security of software, trail ability of Software, ease to modify and improve Software, resistance to crashes feature of Software, advertisement of software, government promotion of software and government willpower to support free software.

9. The responses of users on printed questionnaire and of Google form questionnaire have been converted into SPSS format. In this process, a database file has been developed in SPSS. All the variables have been named in the name column then the questions of the questionnaire have been put in the label field and various choices of the answers to be chosen by the respondent have been put in the value field, and each choice has been given a value like Q1: Which is more costly? Value = 1 proprietary software, Values = 2 free and open source software. All this is known as variable view. The second view is called the data view. Here all the data values have been entered, i.e., 1, 2 and/or so on. At some places 1 means yes and 2 means no.

10. This data has been processed with help of Descriptive frequency test ad one sample T-test using SPSS.

10.3 ANALYSIS OF DATA

Processing of data by applying descriptive frequency test and one sample T-test using SPSS.

10.3.1 TEST CASE 1

Assume H0: Null hypothesis that Users believe that the software of the two categories have equal cost or there is no difference between the costs of two categories of software. H1: Alternative hypothesis: Costs of two categories of software is different. It means one software is more costly than the other.

- **Descriptive Frequency Test:** Table 10.1 displays descriptive statistics providing insight into the pattern of data.

  It is easy to see that mean of number of users who believe that proprietary software are costly are 603 and 206 users believe that FOSS are costly. The figures clearly show that a large number of users know that proprietary software are more costly than FOSS.

TABLE 10.1   Which is More Costly?

		Frequency	Percent	Valid Percent	Cumulative Percent
Valid	1	603	73.9	74.5	74.5
	2	206	25.2	25.5	100.0
	Total	809	99.1	100.0	
Missing	System	7	0.9		
Total		816	100.0		

- **One Sample T-Test:** Table 10.2 shows that output from one sample T-test. The value of t comes to be 16.549. Larger the value of t, the smaller is the probability that the results occurred by chance. Our t value is large. It means results have not come by chance.

TABLE 10.2   One-Sample Test

	Test Value = 1					
	T	Df	Sig. (2-Tailed)	Mean Difference	95% Confidence Interval of the Difference	
	Lower	Upper	Lower	Upper	Lower	Upper
Which is more costly	16.614	808	0.000	0.255	0.22	0.28

- **Df-Degree of Freedom:** This value in our case is 808, which is really close to the total number of participants (816). Missing entries are only 7. It means out of 816, around 7 are not aware of cost of two types of software.
- **Sig. (2 Tailed):** This significance level also called the probability or p-value is 0.000 <0.05, this value is low, very less than 0.05.

It means we reject the null hypothesis that two types of software are equal in cost and alterative hypothesis is proved. The users have opinion that cost of one software category is more than that of the other. In addition, this cost is more for Proprietary software category than FOSS.

10.3.2   TEST CASE 2

Assume H0: Null hypothesis is that Users believe that the software of the two categories has equal freedom to use and reuse or there is no difference between the freedom to use and reuse two categories of software.

H1: Alternative Hypothesis: Freedom to use and reuse of two categories of software is different. It means one software has more freedom to use and reuse than the other.

- **Descriptive Frequency Test:** Table 10.3 displays descriptive statistics providing insight into the pattern of data. It is easy to see that mean of number of users who believe that proprietary software have full freedom to use and reuse are 255 and 538 users believe that FOSS have full freedom to use and reuse. The figures clearly show that a large number of users know that FOSS have more freedom to use and reuse than proprietary software.

TABLE 10.3   Which Category of Software's Provide us Full Freedom to Use and Reuse

	Test Value = 1					
	T	Df	Sig. (2-Tailed)	Mean Difference	95% Confidence Interval of the Difference	
	Lower	Upper	Lower	Upper	Lower	Upper
Which category of software's provide us full freedom to use and reuse	40.877	792	0.000	0.678	0.65	0.71

- **One Sample T-Test:** Table 10.4 shows that output from one sample T-test. The value of t comes to be 40.777. Larger the value of t, the smaller is the probability that the results occurred by chance. Our t value is large. It means results have not come by chance. Entries are only 23. It means out of 816, around 23 are not aware of freedom to use and reuse of two types of software.

TABLE 10.4 One-Sample Test

	Test Value = 1					
	T	**Df**	**Sig. (2-Tailed)**	**Mean Difference**	**95% Confidence Interval of the Difference**	
	Lower	**Upper**	**Lower**	**Upper**	**Lower**	**Upper**
Which category of software's provide us full freedom to use and reuse	40.877	792	0.000	0.678	0.65	0.71

- **Sig. (2 Tailed):** This significance level also called the probability or p-value is 0.000 <0.05, this value is low, very less than 0.05. It means we reject the null hypothesis that two types of software are equal in full freedom to use and reuse and alterative hypothesis is proved. The users have opinion that freedom to use and reuse of one software category is more than that of the other. In addition, this is more for FOSS category than Proprietary software.

10.3.3   TEST CASE 3

Assume H0: Null hypothesis that Users believe that they have no any fear of deskilling from old software, i.e., proprietary software.

H1: Alternative Hypothesis: Users believe that they have fear of deskilling from old software, i.e., proprietary software.

- **Descriptive Frequency Test:** It is easy to see that number of users who have fear of deskilling from old software are 468 and 281 users believe that they have no such fear. The figures clearly show that a large number of users have fear of deskilling from old software.
- **One Sample T-Test:** Table 10.5 shows the output from one sample T-test. The value of t comes to be 21.192. Larger the value of t, the smaller is the probability that the results occurred by chance. Our t value is large. It means results have not come by chance.

- **Df-Degree of Freedom:** This value in our case is 748, which is close to the total number of participants (816). Missing entries are only 67. It means out of 816, around 67 are not aware of such fear.
- **Sig.(2 Tailed):** This significance level also called the probability or p-value is 0.000 <0.05, this value is low, very less than 0.05. It means we reject the null hypothesis that the users have no fear from deskilling, and alterative hypothesis is prove d. The more users have opinion that they fear from deskilling from old software.

TABLE 10.5　One-Sample Test

	Test Value = 1					
	T	Df	Sig. (2-Tailed)	Mean Difference	95% Confidence Interval of the Difference	
	Lower	Upper	Lower	Upper	Lower	Upper
Do you have fear of de-skilling from old softwares	21.192	748	0.000	0.375	0.34	0.41

10.3.4　TEST CASE 4

- **Descriptive Frequency Test:** From Tables 10.6–10.9 displays descriptive statistics providing initial insight into the pattern of data. It is easy to see that number of users who believe that

Reasons for not migrating to open source software system are:

i. Difficult to use = 231 (Table 10.6);
ii. My old files will not open are 243 (Table 10.7);
iii. I am afraid of losing data 284 (Table 10.8);
iv. Do not feel secure 205 (Table 10.9);
v. There are only 53 (Table 10.10) users who have no reason for not migrating to FOSS whereas 706 users (Table 10.10) have one or more reasons out of listed above of not migrating to open source software systems.

Larger the value of t, the smaller is the probability that the results occurred by chance. Our t value is large. It means results have not come by chance.

- **Df-Degree of Freedom:** This value in our case is 758, which is close to the total number of participants (816). Missing entries are only 57. It means out of 816, around 57 are not aware of such fear.
- **Sig.(2 Tailed):** This significance level also called the probability or p-value is 0.000 < 0.05, this value is low, very less than 0.05. It means we reject the null hypothesis that the users have no any reason out of listed above and alterative hypothesis is proved. It means a large number of users have either one or the other reasons out of difficult to use, old files will not open, afraid of losing data; do not feel secure of not migrating to the open source software system.

TABLE 10.6 Reasons for Not Migrating to Open Source Software System are: (1) Difficult to Use

		Frequency	Percent	Valid Percent	Cumulative Percent
Valid	1	231	28.3	30.7	30.7
	2	521	63.8	69.3	100.0
	Total	752	92.2	100.0	
Missing	System	64	7.8		
Total		816	100.0		

TABLE 10.7 Reasons for Not Migrating to Open Source Software System are: (2). My Old Files will Not Open

		Frequency	Percent	Valid Percent	Cumulative Percent
Valid	1	243	29.8	32.6	32.6
	2	502	61.5	67.4	100.0
	Total	745	91.3	100.0	
Missing	System	71	8.7		
Total		816	100.0		

TABLE 10.8 Reasons for Not Migrating to Open Source Software System are: (3) I am Afraid of Losing Data

		Frequency	Percent	Valid Percent	Cumulative Percent
Valid	1	284	34.8	38.1	38.1
	2	462	56.6	61.9	100.0
	Total	746	91.4	100.0	
Missing	System	70	8.6		
Total		816	100.0		

TABLE 10.9 Reasons for Not Migrating to Open Source Software System are: (4) Do Not Feel Secure

		Frequency	Percent	Valid Percent	Cumulative Percent
Valid	1	205	25.1	27.6	27.6
	2	538	65.9	72.4	100.0
	Total	743	91.1	100.0	
Missing	System	73	8.9		
Total		816	100.0		

TABLE 10.10 Reasons for Not Migrating to Open sow-ce Software System are: (5) None

		Frequency	Percent	Valid Percent	Cumulative Percent
Valid	1	53	6.5	7.0	7.0
	2	706	86.5	93.0	100.0
	Total	759	93.0	100.0	
Missing	System	57	7.0		
Total		816	100.0		

TABLE 10.11 One-Sample Test

	Test Value = 1					
	T	Df	Sig. (2-Tailed)	Mean Difference	95% Confidence Interval of the Difference	
	Lower	Upper	Lower	Upper	Lower	Upper
Reasons for not migrating to open source software system are 5. None	100.484	758	0.000	0.930	0.91	0.95

10.4 CONCLUSIONS

According to Test Case 1, the users have opinion that cost of one software category is more than that of the other. In addition, this cost is more for proprietary software category than FOSS. According to Test Case 2, the users have opinion that freedom to use and reuse of one software category is more than that of the other. In addition, this is more for FOSS category than proprietary software. According to Test Case 3, they fear from deskilling from old software, i.e., proprietary software. In addition, this is more for FOSS category than Proprietary software. Finally, according to Test Case 4, a large number of users have either one or the other reasons for not migrating to the open source software system out of difficult to use, old files will not open, afraid of losing data and do not feel secure.

10.5 FINDINGS AND RECOMMENDATIONS

Cost of proprietary software is more than that of free and open source software. This means that if free and open source software is adopted in India, then much of Indian currency will be saved which can be utilized in various development projects for welfare of the public. In addition, India remains in financial crunch and finances saved by adopting FOSS can be much helpful in such scenario. Free and open source software gives the users more freedom to use and reuse which favors the adoption of these software. This aspect again favors the use and adoption of free and open source software. On the other hand users should also report to the development communities. As if the communication between users and the development communities is direct then it will be easy for the developers to make the desired correct updation in the software. To serve the same purpose Google groups and mailing lists are provided by the development communities nowadays. A large number of users have opinion that they fear from deskilling from old software, i.e., the proprietary software. A large number of users have either one or the other reasons of not migrating to the open source software system out of difficult to use, old files will not open, afraid of losing data and not feeling secure. These fears in users call for a strict action from the Government side. Government must address this issue by spreading awareness camps about FOSS and already developed advertisements must be telecast to the public to remove such fears from mindset of the software users of India.

KEYWORDS

- **descriptive frequency test**
- **open source software system**
- **proprietary software**
- **Society for Promotion of IT in Chandigarh**
- **statistical package for social sciences**
- **survey**

REFERENCES

1. Sivaradje, G., Dananjayan, P., & Saraswady, D., (2015). The characteristics of electronic payment schemes: Prospects for the future. *IETE Technical Review*, pp. 197–202.
2. Roumani, Y., Nawankpa, J. K., & Roumani, Y. F., (2017). Adopters trust in enterprise open source vendors: An empirical examination of computer information systems. *The Journal of Systems and Software, 125*, 256–270. The United States.
3. Percio, G., Montesdioca, Z., Carlos, V., & Mac, G., (2015). Measuring user satisfaction with information security practices. *The Journal Computers and Security, 48*, 267–280. Brazil.
4. (2015). *Framework for Adoption of Open Source Software in e-Governance Systems, Version 1.0.* Department of Electronics and Information Technology, Ministry of Communications and Information Technology Government of India, New Delhi.
5. Sreekantan, B. V., (2015). *Role of Electronics in Fundamental Research* (pp. 83–90). Tata Institute of Fundamental Research, IETE Technical Review.
6. (2018). *Introduction to SPSS*. Available at: https://www.uvm.edu/~dhowell/fundamentals7/SPSSManual/SPSSLongerManual/SPSSChapter1.pdf (accessed on 3 November 2020).

CHAPTER 11

Vulnerability of Cyber Threats

ANKIT PRADHAN

Kalinga Institute of Industrial Technology, Deemed to be University, India

ABSTRACT

Technology is advancing faster and widely growing, our world is shifting to its digital form, and a way more advanced way of living our life. today in this scenario mobile phones, laptops, televisions, smart wearables have become a part of our everyday life, and even most of our gadgets in our home is controlled by some or the other form of AI. However, with such high dependency on our digital devices and social media have opened possible targets for various types of cyber crimes like identity thief, fraud, data loss, computer virus, etc. In order to deal with such digital threats and cyber crimes a branch is introduced named "Forensic Science" or digital forensics.

11.1   INTRODUCTION

The forensic branch is the branch of digital crime that deals with the digital thefts and problems faced in the digital world. They work on the information and the evidence found on the various digital devices during an investigation. These information and digital devices mainly belong to suspected criminals or the evidences found at crime scenes.

The whole centre of attention of this field lies on preservation, collection, and analyzation of any digital evidence found during the period of investigation. Digital forensics is used in investigation of various crimes like financial fraud, child pornography, abduction, cyber staking, rape, and other crimes.

Investigators should have a prior knowledge regarding the new disk types, file systems; different components inside a disk, etc., due to the constant development of types and modification of disk, digital forensic investigators need to take intensive training sessions to keep in touch with the current and updated tools.

Digital forensic scientists use a variety of tools, for collecting and analyzing data and digital evidence. It is difficult for one software tool to apprehend sufficient required data, because of which the investigators needs an intensive training later on which it adds on to its learning curve because of which technical knowledge is required, for conducting a deeper investigation, the investigator needs to properly interpret with various results obtained from various tools, and decide their further steps.

11.2 OVERVIEW OF CYBER THREATS

Investigators must retrieve appropriately to refrain from disregarding it in court as indiscriminate evidence, in a cyber security community a malicious act is identified by the continuous attempt to gain access to the system, or the damage being done and what is stolen or the tactics, techniques or the procedures being used.

Since cyber security threats seeks to damage data, illegal data transfer, disrupt someone's digital life, they try to gain unauthorized access, damage, disrupt or steal any information from any unauthorized access to any intellectual property or any other form of sensitive data.

Criminals leave traces that 2010 according to a study released at the end of 2014 by HP and the Ponemon Institute, a research group that studies Internet security [1].

Forensic researchers start their investigation by knowing a background profile of assets beyond the network border and being aware of offline threats, they monitor the domains and IP address ranges, the researchers get enhanced visibility for which they get improved insights into ongoing exploits.

11.3 DIGITAL FORENSICS

The use of scientifically derived and proven methods toward the pres- ervation, collection, validation, identification, analysis, interpretation, documentation, and presentation of digital evidence derived from digital sources for the purpose of facilitating or furthering the reconstruction of events found to be criminal, or helping to anticipate unauthorized actions shown to be disruptive to planned operations [2]. There are different phases of digital forensic branch, given below:

1. **Collection Phase:** In this phase, the forensic investigators gather evidence and collect exact sector level copy of all seized digital media which may contain any potential evidence, the devices from which information can be collected by the investigators hard drives, USB devices, physical RAM, CD_S/DVD_S, and SD cards.

2. **Examination Phase:** In this phase the investigators perform a deep search regarding the collection of evidence and the suspected crime. They identify and locate potential device and other unconventional locations. In this process, the researchers search for obscured data and evidence, but a huge number of data and evidence will be found, data sorting and data reduction can be performed once all the evidences and information are made visible.

3. **Analysis Phase:** In this analysis phase, the researchers go on for a further analysis of the data and evidence collected during the examination phase. The main aspect of this phase is to find and sort the most relevant data and evidence, the investigator finds out the data and evidence using a number of tools and software, in this phase the data that are relevant for the investigation of the case are chosen and collected by the investigator.

4. **Reporting Phase:** It is the last and final phase of the investigation process of any case, in this phase the investigator writes a report regarding the case on which he/she is investigating. The report usually contains an overall summary of the whole examination and the investigation done by the forensic scientist, data collected during the investigation and the relevant information obtained from that collected data and evidence.

11.3.1 *DIGITAL FORENSICS BRANCHES*

On the basis of investigated device, data collected, media, digital artifacts and other factors, digital forensics branch is divided into the following branches:

1. **Computer Forensics:** It is defined by United States Computer Emergency Readiness Team as a multi-disciplinary area that encompasses computer science and law in order to collect and analyze data from computers, networks, and storage devices so that findings can be presented to the court of law as evidence [1].

   Computer forensic branch is used to deal with many problems and crimes like identity thief, financial fraud, data loss, illegal data transfer and other malicious digital acts. The computer forensic branch identifies, analyzes preserves, examines, and works on the investigation and find the ultimate results regarding the case and the crime.

2. **Network Forensics:** This forensic branch mainly focuses on the different computer networks and their traffic in order to detect any illegal or malicious acts undergoing through computer networks. This branch collects data and relevant information and evidence and presents them in the court of law. The data and information obtained in this forensic branch is somewhat volatile and dynamic, that makes the investigation a bit difficult for the investigator, as because when the network traffic is transmitted the data could be lost.

3. **Mobile Device Forensics:** To tackle the crimes and handle all kind of malicious acts related to mobile devices-mobile device forensic branch was introduced. This branch deals with the cellular and smartphones tablets, smart glasses and other handheld devices. Under this branch the handheld devices are investigated and analyzed using various tools and techniques used in mobile device forensics branch.

4. **Database Forensics:** This forensic branch deals with the study of databases and their information about other data. Database forensics needs an attention because of the amount of data required to be investigated of such crimes that databases may be used. For the instance of a financial crimes case, an investigator may need to investigate enormous data in company's databases [1].

5. **Disk Forensics:** This branch gives main attention to the hard disk drives to retrieve forensic information for investigation for any criminal case. The investigator looks at different parts of a hard disk, from where the information can be stored or hidden.

The physical structure of a hard disk is made up of several metal circular disks known as platters, an arm assembly, read-write heads on each arm, and a motor to rotate the disk at a particular speed of 10,000 rpm. The binary data that is written in the hard disk is magnetically. Data is recorded on tracks which are undetectable and have closed centered circles. Each track on the disk is further divided into smaller, more manageable units called sectors.

A sector is the smallest addressable unit on a disk, and was generally 512 bytes in size until January 2011. As of 2011, sector size for all applications using hard disk is standardized by International Disk Drive Equipment and Materials Association (IDEMA) by asking the hard drive industry to move to 4K as the new standard sector size. Since a sector size of 4K is still small this yields too many sectors on a hard disk for the operating system to keep track of (A 500 GB hard disk has over 1 billion sectors!).

11.4 CONCLUSION

The use of this digital forensics is used to solve and keep a track on the malicious act based on the digital crimes can help the technology grow in a very positive way and can reduce the fear of data loss and illegal data transfer. This forensic branch can help humans in streamlining the task and reducing the human errors. This can also make the work of investigators easy for them to work on the evidences and make their investigation easy for them to solve a give their desired result.

KEYWORDS

- **data transfer**
- **digital artifacts**
- **disk forensics**
- **International Disk Drive Equipment and Materials Association**
- **mobile device forensics**
- **standard sector**

REFERENCES

1. www.cs.fsu.edu (accessed on 3 November 2020).
2. Student Paper, Savannah State University.
3. Collingwood Collegiate Institute Technology Department. *Hard Disk Structure*. [online]. http://www.cci-compeng.com/unit5pcarchitecture/5707hddstructure.htm (accessed on 3 November 2020).

CHAPTER 12

Development of Information Communication Systems and Some Social Issues

BALWINDER SINGH BRAR

Department of Applied Sciences, Baba Farid College of Engineering and Technology, Bathinda, Punjab, India

ABSTRACT

Agriculture is the basis-factor; the heavy industry is the leading-factor; and the small-scale industry is the bridging-factor between agriculture and the heavy industry. Advancing on this path, the over-all economic development of our country may be ensured. With fast-developing technology, need for using information-communication systems is increasing in every field. The concept of artificial intelligence (AI) is a part of transmission of information. Advancement of technology is leading to more consumption of energy; hence we are being forced to use more national natural resources. This is also leading to increase in the carbon content in our environment. We must develop such technologies which consume less and less energy and natural energy. We must use facilities based upon developed technologies in a judicious manner. The result of non-judicious use is leading to health problems and diversion from direct and collective dialogue and inter-action resulting in social-isolation among human beings. To control phenomenon of increasing greenhouse effect and to reduce intensity of "energy-crisis," concept of "harvesting energy from atmosphere" should be applied. Question of assessing total combined effect of addition of nutrients present in stubble to soil and evolution of methane gas damaging the plants and raising the greenhouse effect should be seriously debated, correct conclusions must be drawn. There can be development of small power plants using stubble as a fuel, such power plants should be in large numbers so as to consume the enormous amount of stubble available, there may be the factories which can manufacture useful products utilizing stubble as a raw material. Our country is not a developed country. Unemployment is widespread. We have low level of capital-formation. Main thrust of the policy should be to create employment on a continuous and large-scale. For this, industrial, and agricultural development should be carried forward with use of such technology which is labor-intensive. Use of computer-technology everywhere has created unemployment, and decreased the scope of employment. Computers should be used where there is most emergent need. We should produce more and more electricity by using renewable resources and systematically reducing dependence upon non-renewable resources, and discouraging and abandoning the use of nuclear sources of energy.

12.1 INTRODUCTION

The agrarian sector is the basis of industrial development. There is close relation between the development of agrarian sector and small-scale industry. In addition, there is a close relation between the development of small-scale industry and the heavy industry. Thus, there is an organic relation among the development of the three basic aspects of the economy. There should be balanced development of these sectors [1, 2]. The agri-culture is the basis-factor; heavy industry is the leading-factor; and the small-scale industry is the bridging-factor between the agriculture and the heavy industry. Advancing on this path, the overall economic development of our country may be ensured.

Energy is one of the important means of production. Production needs energy of all forms. The energy is being generated from non-renewable as well as from renewable sources. The needs of human beings are increasing continuously. To fulfill these ever-increasing human-needs, there is contin-uous requirement to develop and advance the production-processes, and for all this, there is continuous need to develop and advance science and technology. Use of science and technology in all the production-processes results in energy-consumption on a large-scale. In addition, with the fast-developing technology, the need for using information-communication systems is increasing in every field [3]. As a result, there is increase in the consumption of electric power.

12.2 DEVELOPMENT OF INFORMATION COMMUNICATION SYSTEMS AND ARTIFICIAL INTELLIGENCE (AI)

The role of communication-systems is becoming more and more important. Information theory and coding theory [4–11] have acquired a vital role to improve the faithfulness of transmission of information through information communication systems. This can be felt by the fact of universal use of various codes in the electronics systems, which are used for transmission of information and to execute control in all types of satellites, and military command and control systems. Fire codes [12] were applied to correct errors in magnetic-disc memories and Reed-Solomon (RS) codes have been the main error-correcting codes used in optical-disc memories. Many more error-correcting codes have been developed and are being researched, developed, and implemented.

Encoding reduces the probability of error to a greater extent. If redundancy (i.e., symbols added to source-code to have channel-code) is greater, then error-correcting capacity of the code will be enhanced. On the other hand, if redundancy is more, then the energy per channel-symbol will be more reduced as compared to the energy per data-symbol. Therefore, symbol-energy will be decreased and this will lead the demodulator to commit more errors. Therefore, with greater redundancy, there will be enhancement of error-correcting capacity of the code; but there will also be chance of more errors.

Communication [13–15] is the process in which link between two points is established for exchanging information. In wireless communication, medium of transmission of information is the open-space, and transmission is by electro-magnetic waves. From the transmitter, radio waves are radiated in open-space through antenna. At the receiver, another antenna intercepts the radio waves. The signal containing information is known as modulating signal or baseband signal. Modulation is a process in which some characteristic of a signal, called carrier, is changed in accordance with instantaneous value of another signal, called modulating signal or baseband signal. Therefore, in modulation, a message-signal is placed over carrier-signal so that message-signal (modulating-signal or baseband-signal) becomes suitable for transmission over longer distances. The frequency of carrier-signal is greater than the frequency of message-signal (modulating-signal or baseband-signal). The signal resulting from the process of modulation is known as modulated signal.

From transmitter, modulated-signal is transmitted and it goes through the channel, and it is received by the receiver. At receiver, to reconstruct the original signal (baseband signal), the modulated-signal will have to pass through a reverse-process, called demodulation. Therefore, demodulation is a process in which message-signal is separated from the modulated-signal.

The energy consumption [16] by communication-systems can be seen more clearly by the fact that both encoding and modulation are integrated (similarly both decoding and demodulation are integrated). So, in light of hardware implementation of encoding and modulation (and implementation of decoding and demodulation), the energy consumption is of the following kinds: (i) For channel encoding and decoding, computational-energy is needed; (ii) For modulator and demodulator, circuit-energy is needed; (iii) For transmitting redundant bits, signal-energy, i.e., radio-energy is required.

If the distances are small, then the energy spent in the transmission and receiver circuits and in computation will be comparable to signal-energy. If the distances are large, then along with the energy spent in the transmission and receiver circuits and in computation, the signal-energy will be more pronounced.

To achieve the aim of improvement of error-correcting capability of the code, we are to increase the redundancy, with increase of the redundancy, there may be occurring of more errors. Therefore, there will be some limit to the redundancy, and hence some limit to the error-correcting capability of the code.

There is also the other aspect. To have the error-correcting capability of the code, we involve the redundancy, and to have the improvement of error-correcting capability of the code, we are to increase the redundancy, but with increase of the redundancy, the length of the code will increase. As a result, the speed of the transmission of the information from source to the destination will be decreased. So, more time and hence more electricity will be needed to send the information.

As a result of development and large-scale use of information communication systems, more electricity is required for the transmission of information. So, problems of energy-crisis will become more and more serious.

Concept of artificial intelligence (AI) is a part of transmission of information. As a concept, AI means that a machine mimics cognitive functions which humans associate with other human minds, such as learning and problem solving. Taking from the management literature, Kaplan and Haenlein explained AI as constituting three different types of systems: (i) analytical meaning that it has those characteristics which are consistent with cognitive intelligence generating cognitive representation of the world and using learning based on past experience to estimate the future decisions; (ii) human-inspired meaning that it has elements from cognitive as well as emotional intelligence and understanding, considering them in their decision-making; and (iii) humanized AI. It shows characteristics of all types of capabilities such as cognitive, emotional, and social intelligence, being able to be self-conscious and self-aware in interactions with one another. Aims of AI research include reasoning, knowledge, representation, planning, learning, natural language processing, perception, and the ability to move and manipulate objects. The various approaches such as statistical-methods, computational-intelligence, mathematical-optimization, probability and economics, computer science, and engineering, information-engineering, mathematics, psychology, linguistics, philosophy, etc., are employed [17].

The field of AI claims that human intelligence may be so precisely described that machine can be made to simulate it. This project the questions concerning nature of the mind and ethics of creating artificial beings endowed with human-like intelligence. There is a view that unlike previous technological revolutions, AI will create mass unemployment [17].

The concept of AI is basically related to and is part of the phenomenon of the transmission of information and synthesizing the information to arrive at certain estimated solutions of the problems. Therefore, all problems related to transmission of information are present in the domain of AI.

Firstly, we should use and develop AI as we are using other technologies. Secondly, it cannot be made as an alternative to the human-brain and human-mind (human thinking). So AI should be utilized as an important supplementary tool. We should use AI, but not depend upon it. We should rely on human-brain and human-mind (human thinking). Thirdly, the problems of unemployment by use of this concept on any-scale should be anticipated beforehand and pre-planned and solved. Fourthly, by use of this concept on any-scale, the consumption of electricity will increase further, and this will add to the gravity of energy-crisis.

12.3 MORE AND MORE ENERGY IS REQUIRED AS WELL AS PRODUCED

During modulation, since the low frequency signal is transformed to higher frequency spectrum, and because:

$$E = h\,\upsilon \qquad\qquad (1)$$

where; h is the Plank's constant having value equal to $6.62607015 \times 10^{-34}$J.s.

i.e., $\propto$ Energy frequency (υ).

Therefore, in the process of transmission of information, so much energy is involved, i.e., more, and more energy is required as well as produced. As a result, more, and more heat will be produced. Therefore, we have two consequences, one is there is increasing consumption of electrical energy, and two, there will be increasing evolution of heat energy. The first aspect will lead to deepen the already existing energy-crisis, and the second aspect will certainly facilitate the rising the atmospheric temperature, which will have two-fold far-reaching effect: To increase greenhouse effect thereby increasing all the ill-effects of increased greenhouse effect; To damage the human-health in the form of damage to the health of all those people who handle and work with the concerned devices and all the people living in the areas through which the antennas are installed and all the space through which the high frequency-signals are going in all the directions.

To control the phenomenon of increasing greenhouse effect and hence to control its increasing ill-effects, the concept of "harvesting energy from atmosphere" should be understood, scientific, and engineering communities should be made more aware, and this concept should be developed and applied. The basic point in this is that the heat energy and its transformed forms present in the atmosphere must be tapped and used to run various devices. For this, the methods and technologies to tap this energy will have to be developed. In addition, we will have to develop the various devices in such a way so that these devices may be able to utilize the increasing heat energy and its various transformed forms in an automatic manner.

12.4   IMPLICATIONS OF RUTHLESS USE OF FACILITIES BASED UPON DEVELOPED TECHNOLOGIES

12.4.1   HEALTH PROBLEMS

We must use the facilities based upon developed technologies not in a ruthless manner but in a judicious manner. However, this is not happening. The result of this is also leading to health problems, especially of eyes due to electro-magnetic radiations [18], and diversion from direct dialogue and collective inter-action leading to social-isolation among human beings.

There are so many ill effects of computer-technology on human-health. While working on a computer, there is continuous strain on the eyes. CVS (computer vision syndrome) has been there in IT-Industry for many years, the symptoms of which are: eyestrain, blurry-vision, dry-eyes, fatigue, and double vision. When a person sits to work on a computer for 3–4 hours continuously without break, then: (i) blinking-rate of the eyes is reduced to a considerable extent, i.e., 2–3 times per minute as compared to normal person's blinking-rate of 10–15 times per minute, the reduced blinking-rate results in reducing wetting of the eyes which further leads to lowering the supply of oxygen to the eyes; (ii) focusing-problem meaning thereby that eyes become unable to focus smoothly and easily on a particular object even long after the completion of work; (iii) eye-irritation is caused. In addition, there are results in the form of headache, muscle-pain, joint-pain, neck-pain, and shoulder-pain due to continuous-sitting to work on the computer. The eyes are strained severely, which further ill affects the various parts of the human-body. If there are serious vision-problems, then one has to sit in a wrong-posture to compensate these vision-problems, and this will definitely result in developing neck-pain and back-pain. In addition, working on computer results in repeating same movements, again and again, using the same muscle groups in hands, arms, and shoulders, all this leading to a repetitive-stress-injury. All the electronic-devices, like computers, mobiles, TV, etc., emit electro-magnetic-radiations (EMR) which have heating-effect. As compared to other tissues, cornea has fewer vessels; hence cornea is more likely to be affected by heating-effect. This may cause cataracts. In addition, there is evidence that these radiations increase the rate of skin-rashes and miscarriages.

12.4.2 UNEMPLOYMENT PROBLEMS

Our country is not a developed country. Here unemployment is widespread, not only in vast rural areas, but also among semi-skilled, skilled workers, and technicians and engineers. On the other hand, we are poor in capital and have low level of capital-formation. Therefore, here main thrust of the policy should be to create employment on a continuous and large-scale. In this light, we see that use of computers every-where has created unemployment, and also decreased the scope of employment. Therefore, computers should be used where there is the most emergent need of these like in the research-work, in military command and control, in guiding the space-vehicles, in mining, in oil-and-gas-exploration, in handling the sophisticated machines, and so on.

12.4.3   *DIVERSION FROM DIRECT DIALOGUE AND INTER-ACTION LEADING TO SOCIAL-ISOLATION*

The use of communication-systems and communication-technology is being used ruthlessly leading to become a full-fledged "Internet-addiction." Use of various so-called social-media has become universal. People say that they have thousands of friends on Facebook, but they have none in real-life. How can it be said to be the social-media? Use of this so-called "social-media" has been leading to make human beings as isolated-atoms, divorced from real situations and real problems as a whole. In families, most of the members are all the time engrossed in the mobiles, computers, and internet and mutual-interaction on the family-level is going to decrease further and further. Similarly, at the locality-level, social-interaction is on decrease. As a result, diversion from direct and collective dialogue among people is being promoted and hence the social-interaction is being damaged. The basic path of seeking knowledge and implementation of the gained knowledge and generalization of the human-experience is of the direct and collective dialogue and the social-interaction, and it is being eroded by using internet in a ruthless manner. It does not mean that we should not use the internet in different forms. We should use it in a judicious manner, taking it as a helping-hand and giving it a supplementary-role and not as a substitute of direct and collective dialogue and social-interaction [19, 20].

12.5   SOURCES OF ENERGY

12.5.1   *QUESTION OF ENERGY*

We must have to analyze the direction of generating and developing the energy. On the world-level, it is being produced in a big way from non-renewable sources like coal, petroleum, natural gas, and nuclear sources (nearly 90%) and energy from renewable sources is very small (nearly 10%) [21]. As a result, carbon-content is increasing tremendously. In India at present, the major sources of electrical energy are fossil fuels and water [22].

In the last many decades, there was attempt to switch from traditional non-renewable sources to nuclear sources. The capital cost of a nuclear power plant is too high [22]. Nuclear sources are themselves non-renewable sources. In addition, the accidents in nuclear power plants in America, Russia, and Japan have projected the very serious hazardous nature of producing nuclear energy. Therefore, in recent times, use, and development of nuclear sources of energy is being discouraged and abandoned on the world-level, and it should be.

As far as renewable sources are concerned, there was sound development of energy from water (hydro-electricity), and there is yet the further scope of development of this (also at micro-level). In India, the hydroelectric resources are estimated at 85,000 MW, out of which only 23% has been developed so far [22]. The development of energy from air and sunlight is to a small extent. On the whole, production of energy from renewable sources should become the main stay.

We should produce more and more electricity by using renewable resources and systematically reducing dependence upon the generation of electricity from non-renewable resources, and discouraging and abandoning the use of nuclear sources of energy.

12.5.2 PROBLEM OF STUBBLE-BURNING

There is an acute problem of stubble burning and the resulting problem of air-pollution leading to various ailments in our country, especially in north India. There are some methods which are being suggested to deal with it. Some say that using some machines; the stubble should be put back into the soil to raise the fertility of the soil. There are two aspects concerning this suggestion. First are that such machines and the technology will raise the cost of agriculture which is already high and beyond the capacity of the peasantry. The peasants are already living in the conditions of poverty and in the stranglehold of banking and usury debt, and as a consequence of that many peasants are being pushed to commit suicides. So to say to the peasants to withstand this extra cost of applying the technology to put back stubble into the soil would be a cruel joke on them. Second aspect of putting back stubble into the soil will not necessarily raise the fertility of the soil. Although by the bio-degradability of the stubble in the soil there will be addition of nutrients present in the stubble to the soil, but in the same process of bio-degradability, there will also be evolution of methane gas (CH_4), which will damage the plants and also mix with the atmosphere leading to raise greenhouse effect. So, the question of assessing the total combined effect of addition of nutrients present in the stubble to the soil and evolution of methane gas (CH_4) damaging the plants and raising green-house effect, is there, and it should be seriously debated and vital conclusions must be drawn whether the sum-total of these two opposite effects is a positive or negative. The third aspect is that there can be development of small power plants in which electricity should be generated by using stubble as a fuel, and such power plants should be in large numbers so as to consume the enormous amount of available stubble. Note that the stubble is a renewable resource. In addition, there may be the factories which can manufacture useful products utilizing the stubble as a raw material. With such an effort, problem of stubble burning on mass scale may be solved; there will be generation of electric-power; there will be manufacture of useful products; there will be income to the peasants by selling the stubble.

12.6 CONCLUSIONS

1. Advancement of technology is leading to more and more consumption of energy and to fulfill this increasing need of energy, we are being forced to use more and more national natural resources. In addition, this is also leading to increase in the carbon content in our environment.

2. We must develop such technologies [23, 24] which consume less and less energy. For example, green computing [3], where green computing is the study and practice of environmentally sustainable computing (IT). There should be study how to perform the jobs minimizing the power-consumption.

3. We must develop such technologies which consume the natural energy.

4. We must use the facilities based upon developed technologies in a judicious manner. The result of ruthlessness is leading to Health problems, especially of eyes due to electro-magnetic radiations [18], and diversion from direct and collective dialogue and inter-action leading to social-isolation among human beings.

5. To control the phenomenon of increasing greenhouse effect and hence to control its increasing ill effects, and also to reduce the

intensity of "energy-crisis," the concept of "harvesting energy from atmosphere" should be applied.

6. Question of assessing the total combined effect of addition of nutrients present in the stubble (a renewable resource) to the soil and evolution of methane gas (CH_4) damaging the plants and raising the greenhouse effect, should be seriously debated and correct conclusions must be drawn. There can be development of small power plants in which electricity should be generated by using stubble as a fuel. In addition, there may be the factories which can manufacture useful products utilizing the stubble as a raw material.

7. Our country is not a developed country. Unemployment is widespread. We have low level of capital formation. Here main thrust of the policy should be to create employment on a large-scale. For this, industrial, and agricultural development should be carried forward with the use of such technology which is labor-intensive. Use of computer-technology every-where has created unemployment, and also decreased the scope of employment. Therefore, computer-technology should be used where there is the most emergent need like in the research-work, in military command and control, in guiding the space-vehicles, in mining, in oil-and-gas-exploration, in handling the sophisticated machines, and so on, and hence there should be a comprehensive review of the whole system of utilization of computer-technology and necessary steps should be taken up.

KEYWORDS

- artificial intelligence
- carbon-content
- energy and energy-crisis
- harvesting energy from atmosphere
- health problems
- information-communication-systems
- judicious
- labor-intensive

- **national natural resources**
- **renewable resources**
- **social-isolation**
- **transmission of information**
- **unemployment**

REFERENCES

1. Ruddar, D., & Sundharain, K. P. M., (2007). *Indian Economy*. ISBN: 81-219-0298-3, S. Chand & Company, New Delhi.
2. Lekhi, R. K., Singh, & Joginder, S., (2006). *Agricultural Economics* (5th edn). Kalyani Publishers, Ludhiana.
3. Prashant, G., Sanjay, B., & Deepali, (2014). Green computing. *Research Cell: An International Journal of Engineering Sciences, 3*. ISSN: 2229-6913 (Print), ISSN: 2320-0332 (Online).
4. Hartley, R. V. L., (1928). Transmission of information. *Journal "Bell System Tech.," 7*(3), 535–563.
5. Shannon, C. E., (1948). The mathematical theory of communication. *Bell System Tech. J., 27*, 379–423.
6. Fano, R. M., (1949 & 1950). The transmission of information. *Tech. Rept., 65, Rept. 149*. Mass. Inst. Technol., Research Lab. Electronics.
7. Fazlollah, M. R., (1961). *An Introduction to Information Theory*. McGraw-Hill, New York.
8. Hill, R., (1986). *A First Course in Coding Theory*. Oxford University Press.
9. Stevan, R., (1992). *Coding and Information Theory*. Springer-Verlag, New York, Berlin, Heidelberg, London, Paris, Tokyo, Hong Kong, Barcelona, Budapest.
10. Vinocha, O. P., & Bhullar, J. S., (2005). A note on information theory and cryptogrphy. *8th National Conference of Indian Society of Information Theory and Applications* (pp.79–82). Malout (Punjab).
11. Brar, B. S., (2010). *Foundations of Information Theory* (p. 103). National Conference on Smart Electronic and Engineering Materials 2010 (SEEM`s10), organized by Baba Farid College of Engineering and Technology, Bathinda on 5th & 6th March, 2010. The Abstract of the paper was published in Book of Abstracts of the Conference.
12. Fire, P., (1959). *A Class of Multiple-Error-Correcting Binary Codes for Non-Independent Errors.* Thesis/dissertation: e-Book, Department of Electrical Engineering, Stanford University.
13. Simon, H., (2001). *Communication Systems*. Eighth Reprint, John Willey & Sons, Inc., New Delhi.
14. Simon, H., (1988*). Digital Communications*. John Willey & Sons, ISBN: 978-81-265-0824-2, authorized reprint by Willey India (P) Ltd., New Delhi.

15. Singh, R. P., & Sapre, S. D., (2008). *Communication Systems (Analog and Digital).* Fourth Reprint, Tata McGraw-Hill Publishing Company Limited, New Delhi.

16. Ranjan, B., (2008). *Information Theory, Coding and Cryptography.* Tata McGraw-Hill Publishing Company Limited, New Delhi.

17. Online: https://en.wikipedia.org/wiki/Artificial_intelligence (accessed on 3 November 2020).

18. Brar, B. S., (2014). Computer science and some problems. *An International Journal of Engineering Sciences, 3*, 37–40. ISSN: 2229-6913, e-ISSN: 2320-0332.

19. Nikhila, P. D., (2015). *Internet Addiction Leading to Psychological Disorders: Experts.* The Tribune.

20. Rajesh, G., (2015). *The Myth and Virtual Reality of Social Networking.* The Tribune.

21. Rai, G. D., (2014). *Non-Conventional Energy Sources* (5th edn., p. 5). 2011, 12th Reprint, 2014, Khanna Publishers, New Delhi.

22. Gupta, B. R., (2013). *Generation of Electrical Energy*. Published by Eurasia Publishing House (Pvt.) Ltd, New Delhi, ISBN: 81-219-0102-2.

23. Brar, B. S., (2015). *Pro-People, Pro-National Technology, Production-Systems and Management-Models* (pp. 68–76). 1st National Conference, on "Human Values & Professional Ethics," organized by Department of Mentoring & Value Inculcation, Chandigarh University, and full paper was published in Conference Proceedings (ISBN: 978-93-84468-44-6).

24. Brar, B. S., (2016). *Need for Supreme Prominence to the Development of Railways* (Vol. 5, No. 1, pp. 40–45, 53, 54). 4th International Conference on "Advancements and Futuristic Trends in Mechanical and Materials Engineering", organized by Baba Farid College of Engineering and Technology, Bathinda, Punjab, under the aegis of Society of Materials and Mechanical Engineers (SOMME), sponsored by MRS State Technical University, Bathinda (Punjab), DST-SERB, and Indian Society For Technical Education (ISTE), and published on Souvenir (Proceedings) of the conference, and also published in "Asian Review of Mechanical Engineering" [An International Peer-Reviewed Journal of Mechanical Engineering], ISSN: 2249-6289.

CHAPTER 13

Role of Digital Technology in Collaborative Learning: An Analysis

DEEPANJALI MISHRA

School of Humanities, Kalinga Institute of Industrial Technology, Deemed to be University, Bhubaneswar, Odisha, India

ABSTRACT

Collaborative learning must not be confused with individual learning because considered to be different from each other. It is a form of learning which is more important for versatile productivity in a creative and atmosphere of mass accountability that gives rise to more benefit. This mode, which takes place through interaction and exchanging ideas among students of different learning levels, benefit from this method of learning strategy and it promotes a very positive and amicable causing harmony among all the students. This chapter reveals the learning in positive interdependence of a face to face sitting setting. The multiple outcomes studied so far throw light on three major aspects: positive relationship, achievement productivity, and psychological health as the advantages of collaborative learning. It is a learning atmosphere in togetherness to accomplish certain objectives. One's excellence is mutually brought to passerby others performance and achievement. The knowledge, skill, and resources of an individual are seen multiplied copiously in a group presentation, in comparison to an individual endeavor. Pluralism and diversity among different talents authorize a positive thwack on every partaker on the basis of social psychology and classroom interactions. The positive impact of collaborative learning in various studies is appraised by the improvements of personal attributes ignoring irrelevant failures and drawbacks in a leading-edge ambience. "Individual learning in an isolated atmosphere may demonstrate a paucity of creative, cognitive, moral reasoning, on the contrary, be supplemented by a variety of cultural and ethnic backgrounds, cooperation supplanting competition. Therefore, this chapter proposes a critical analysis of the concept of collaborative methodology, challenges subjugated in this domain and how could it be overcome.

13.1 INTRODUCTION

Collaboration is a promising mode of human engagement and that has collaborative learning is an educational approach to teaching and learning that brings in the inculcation of learners' endeavor to accomplish a task, resolve a complication or innovate something new. The benefits of collaborative learning are proliferated with sundry advantages of social, psychological, academic, and assessments of the learners. Here the children come together to deal among themselves, exchanging their impressions and intentions through their multiplied abilities, resources, skills, and contributions. As Cohen, S., and Willis, T stated, It creates a stronger social support system. "There takes place a sharing of potentiality and acceptance of responsibility among different members for the group's common action." The underlying essence of collaborative learning is based upon concurrent manifestation of propensities through cooperation by group members, in contrast to competition in which individuals bet other group members. The three possibilities the learners have in a collaborative learning may be; either they promote by collaboration, obstruct by competition or remain neutral to each other by individualistically effort to attain success. This can happen anywhere in a gathering of common goal. The technique of collaborative learning is applicable in the classroom, with community groups, within families, as a way of living and dealing with other people. "From the point of cognitivism in psychology or functionalism in philosophy it can be learnt that collaborative learning ambience is much more impactive on the children than an individual learning situation." For a collaborative learning, efforts need to be more productive than competitive or individualistic methods become a twenty-first-century trend. As per Austin comments, "the growing demand of 'thinking together' and 'working together' on various critical issues has been rampant." The shift away has caused to stress on from individual attempts to teamwork and from autonomy to community. The idea of collaborative learning, the grouping, and pairing of learners intending to achieve a learning goal has been widely researched and championed. The term collaborative learning refers to an instruction method in which learners at various performance levels work together in small groups intending a common goal. The learners take responsibility for one another's learning as well as their own. Thus, the performance of one learner helps the other's performance.

Collaborative learning was originated from Lev Vegtosky's zone of proximal development theory. In this theory the significance of learning through interaction and communication with others has been highlighted than the learning through independent effort. The gap between things a child can do and can't do is made up by collaborative learning. According to Gokhale [5], "individuals are able to achieve higher levels of learning and retain more information when they work in a group rather than working individually. Collaborative learning is useful for both the facilitator and the learner as it accelerates the learning through its motivating atmosphere." Both being group learning mechanism, collaborative learning differs from cooperative learning by having a mutual, coordinated effort of members unlike a shared responsibility imposed on each member in the latter. In other words, while collaborative learning deals with the construction of interaction; cooperative learning deals with its philosophy. Most of the time, collaborative learning is seen being used as an umbrella term for a variety of approaches through interdependent learning activities. Many have found this to be beneficial in helping students learn effectively and efficiently than by learning independently, thus having positive attitude about learning and being growing into more engaged and thoughtful learners.

When compared to more traditional methods where students non-interactively receive information from a teacher, in this method of learning, lower-ability students work better in mixed groups and medium-ability students do better in homogeneous groups. "For higher ability students, of course, it may not be so useful. Research says that discussion-based practices, improved comprehension of the text and critical-thinking skills for students across ethnic and socioeconomic backgrounds are found enhanced through this learning." The popularity of collaborative learning in the classroom has increased over the last decade. Web technologies create learner-centered learning environments with individual inducement. Collaboration becomes highly necessary day by day. There is implication for a lot of future work, in order to have collaborative learning highly effective. Some of the unsolved problems that may be identified are cultural diversity, and accordingly a lack of awareness of cultural norms, geographical distance and time zone differences, and member seclusion in virtual set, generation intermissions and age contrasts in the acceptance of collaboration tools, lack of aids and appliances for learners, lack of learners' awareness about efficacious collaboration processes and schedule, lack of learners' application skills and knowledge about collaboration tools. It is undoubtedly significant to consider the interactive processes among pupils, but the most critical part is the construction of new knowledge obtained through joint work.

13.2   ADVANTAGES OF COLLABORATIVE LEARNING

Collaborative learning provides scope for promoting interaction aiming the group goal through individual accountability and personal responsibility. The interpersonal relationship and group skills become distinctively improved. In a collaborative learning milieu, learners burgeon responsibility for one another socially the learners fit themselves to understand diversity and cooperate overtly. Other than this psychologically they become calm and stable keeping all anxieties aside fostering their self-esteem and expressing in an ensemble. Academically they get their critical thinking and problem solving skill inflated. Their active participation in the learning process improves the class result as a whole. They grow a positive slant towards the teacher. It helps to develop learning communities within classes and institutions. Since students get actively involved in interacting with each other regularly in an instructed mode, they get scope to understand their differences and learn how to set right social problems which may appear (It creates a stronger social support system. Collaborative learning induces learners' motivation in a specific curriculum. Through alteration of student teacher assessment techniques it makes learning more open and wide. Collaborative learning is a free and independent learning. The mandate imposition and restrictive guidelines may make the learning virtually nonexistent. The ambience with ample scope for reflection and autonomy paves the way of collaborative learning convenience. As rightly stated by Cohen, collaborative learning ameliorates social interaction skills. Collaborative learning medium provides students with opportunities to analyze, synthesize, evaluate, and create ideas cooperatively. The informal setting facilitates discussion and interaction. This group interaction helps students to learn from each other's scholarship, skills, and experiences. The students go beyond mere statements of opinion by giving reasons for their judgments and reflecting upon the criteria employed in making these judgments. Thus, each opinion is subject to careful scrutiny in a group.

The focus on social and emotional aspects as benefits of collaborative learning is as important as the cognitive part. The most interesting part of this learning process is the decreased amount of anxiety and nervousness for solving a problem or taking a decision is seen paramount due to shared responsibility. Sense of humor also plays a significant role to reduce anxiety. In a happy emotion the learners take active part to opine their own critical appreciation through group under the facilitation of the teacher. This involves constructing and controlling meaningful learning experiences and stimulating students' thinking through real world problems.

13.3   ANALYSIS OF AN EFFECTIVE COLLABORATION TEACHING

To make the art of collaboration effective, the educators have to take thoughtful considerations for its smooth execution. In many good schools collaboration schedules are made in order to have a planned sharing of the teachers for a common achievement of the school. Successful teaching partnership stimulates the best professional skills and practices for the holistic development of the children through the implementation of teaching methodologies.

Collaboration is a wonderful teaching tool. Teachers here find the scope to evaluate and differentiate instruction for students more distinctly and they can learn new instructional craftsmanship from one another to expand their teaching repertory. Cooperative teaching experiences also provide mutual support and assistance for planning and implementing lessons, assessing students' progress, sharing professional concerns, and addressing students' learning needs. Most importantly, teaming allows more opportunities for students to understand and connect with content thereby maximizing individual learning potential. Considering different ways of team teaching which can be used effectively in the classroom, it is anonymously accepted as a popular instructional model. Collaborative teaching allows teachers to impart information to a broader range of learners using approaches that spark students' imaginations while supporting individual learning.

13.4   COLLABORATIVE LEARNING AS A QUALITY IN TEACHERS

Good teachers are no more than good learners. They are emerging practitioners. They venture to grow their knowledge through the input of the expertise of their compeers locally and globally, physically or virtually by social media. The professional growth, innovations in teaching, the students' achievement are laid on the teachers' interconnection and intercommunication. "The monthly sharing meetings, weekly staff meetings, professional learning community (PLC) with other members such as department head, content experts, subject specific resource persons, are held on the purpose. The advice, suggestions, reciprocity of thoughts and plans impact and promote positively each other." Veteran teachers can render as a worldly-wise professionals for the novice and the newbie teachers flicker exhilaration among the veterans. In this connection at least a like-minded colleague is highly required even through technology for the multiplicity of the productivity.

Different variables of collaborative learning process may include group composition such as homogenous or heterogeneous groups, the selection, and size of groups, role of teacher as a facilitator, structure of groups, preferences in concern to gender and ethnicity, learning styles and strategies, psychoanalysis of the group discussion, etc., are to be minutely taken into consideration.

13.5 WAYS OF TEACHING COLLABORATIONS

The term collaborative learning refers to an instruction method during which learners at various performance levels work together in small groups towards a standard goal. The learners are liable for one another's learning also as their own. Thus, the success of 1 learner helps other students to achieve success. Students can get entangled in developing curriculum and sophistication procedures (Kort, 1992). Students are often asked to assess themselves, their groups, and sophistication procedures (Meier and Panitz, 1996). The high level of interaction and interdependence among group members results in deep instead of surface learning (Entwistle and Tait, 1993). Collaborative learning is student-centered, resu lting in a stress on learning also as teaching and to more student ownership of responsibility for that learning (Lowman, 1987). Collaborative teaching styles are nowadays utilized in wide varieties. Such as:

1. **Leading, Observing, and Assisting:** After the new content is presented by the lead teacher, the co-lead teachers lead the discussion, observe it, and assist wherever there's a requirement of interference.

2. **Teaching and Re-Teaching:** The lead teacher delivers new material through different activities, while the co-instructor reviews previous information and skills for extended retention purposes.
3. **Simultaneous Teaching:** The category is split into smaller groups and teachers present an equivalent material to different groups at an equivalent time.
4. **Instructional Stations:** Students rotate between several stations to receive new instructions and participate on different activities monitored by teachers.
5. **Supplementary Teaching:** During this style one teacher deals and provides instruction to the bulk of learners, the opposite takes a little group aside to figure on different basic instructional goals of literacy skills.
6. **Co-Teaching Rotation:** Here the teachers present new information rotating between presentations and scaffold roles during the lesson.

13.6 USE OF MULTIMEDIA IN COLLABORATIVE LEARNING

Technology plays an important role in collaborative learning. The last 10 years' details show that internet has provided a good space for groups to communicate adequately. Virtual learning groups have been useful to allow people to communicate with far distant people. Research has been conducted on how technology has helped to increase the potential of collaborative learning, to build an online learning environment model but since this research was conducted the Internet has fattened extensively and the new software has changed these means of communication.

Here are some examples of the way technology becomes increasingly integrated with collaborative learning. Collaborative-networked learning occurs via electronic dialogue between self-directed co-learners and learners and experts. Learners are directed towards a common purpose and are accountable to one another for their success [9]. Collaborative-networked learning occurs in interactive groups where participants actively communicate and negotiate learning with each other within a contextual framework which may be facilitated by an online coach, export, and mentor or group leader.

Computer-supported collaborative learning (CSCL) may be a relatively new educational paradigm within collaborative learning which uses technology during a learning environment to assist mediate and support group interactions during a collaborative learning context. CSCL systems use technology to regulate and monitor interactions, to manage tasks, rules, and roles, and to mediate the acquisition of latest knowledge. Wikipedia is an example of how collaborative learning tools are extremely beneficial in both the classroom and workplace setting. They're ready to change supported how groups think and are ready to form into a coherent idea supported the requirements of the Wikipedia user. Collaborative learning in virtual worlds by its nature provides a superb opportunity for collaborative learning. Educational organizations can employ groups to urge the work done. "Collaborative models in group work paradigm are often guessed from the massive scale professionals' interest within the modern trend of business. Same thing being applicable to educational sphere, the advents of this learning approach is to be analyzed. Extensive research thereon either face-to-face or computer supported has thrived the skate-holders within the past ten years. This approach to education being multidisciplinary and non-competitive in nature, it sensibly embraces a mess of theoretical and practical accounts of opportunities and problems. This learning concept has been vastly researched and advocated as an instructional method of exchanging ideas. The shared learning gives students a chance to interact in discussion, take responsibility for his or her own learning, and thus they grow into critical thinkers. Most of the studies in collaborative learning are wiped out non-technical disciplines. Students are exposed to think creatively, solve problems, and make decisions because the education network among the learners, functions like both as series and parallel connection of electricity during a circuit. As Slavin (1989) says, for effective collaborative learning, two important things must be "group goals" and "individual accountability." As explained, students are capable of exhibiting at higher intellectual levels when asked to figure in collaborative situations as compared to when asked to figure individually. Group miscellany in terms of experience and knowledge contributes positively to the training process [10]. Bruner [1] discovers that cooperative learning methods improve problem-solving strategies because the scholars are tackled with different demonstrations of the given situation. The group network enables the learners to internalize both exterior knowledge and interior critical thinking skills and to convert them into contrivances for intellectual functioning. Today's kids of 'digital generation' and students of 'smart class' are quite competent with the technologies of varied authoring programs; like Microsoft Word, PowerPoint, Microsoft's Photo Story 3,

Windows XP, Apple Computer's iMovie, camera, video-cam recorder and scanner, computer microphones and digital voice recorders to form their collaborative learning lively hundred times, only needing the right channelization of their potentialities. Use of audio visual aids and group presentations through it makes the training interesting, long lasting and enlightens the upcoming generations with extensive learning outcomes. About collaborative learning it's proved of resulting in self-management Students are skilled to be prepared to finish the assignments and work together within their groups understanding the matter that they decide to contribute to their groups [11]. They're, moreover, given time to process group behaviors like examining with one another to make sure homework assignments not only to be completed but also understood. These interactions help students learn self-management techniques. As Cooper says, Collaborative learning provides the teacher with many opportunities to watch students interacting, explaining their reasoning, asking questions, and discussing their ideas and ideas. These are more comprehensive assessment methods than counting on written exams only (Cross and Angelo, 1993). Johnsons (1990) further delineates, during a learning situation, student goal achievements are positively corresponded; students perceive that they will reach learning goals if the opposite students within the learning group also outstretch their goals. During this way, students solicit outcomes that are beneficial to all or any those with whom they're collaboratively linked. When individuals wedged they're more likely to surrender, but groups are often likely to seek out ways to stay on. Collaborative learning provides many opportunities for alternate sorts of student assessment (Panitz and Panitz, 1996). Collaborative learning minimizes classroom anxiety, because it always creates new and unfamiliar situations for the scholars [12]. The scholars observe that the teacher has the power to gauge how students think also as what they know. Through the interactions in the collaborative learning process, with students during each class, the teacher achieves a far better understanding of every student's learning style and strategy, how he/she performs and a chance is made whereby the teacher may provide extra guidance and counseling to the scholars.

13.7 CONCLUSION

Collaborative learning compared with competitive and individualistic efforts, has numerous benefits and typically results in higher achievement and greater productivity, more caring, supportive, and committed relationships; and greater psychological health, social competence, and self-esteem. Therefore, Collaboration is a philosophy of interaction and personal lifestyle where individuals are responsible for their actions, including learning and respect the abilities and contributions of their peers. In all situations where people come together in groups, it suggests a way of dealing with people which respects and highlights individual group members' abilities and contributions. There is a sharing of authority and acceptance of responsibility among group members for the groups' actions. The underlying premise of collaborative learning is based upon consensus building through cooperation by group members, in contrast to competition in which individuals best other group members. A natural tendency to socialize with the students on a professional level is created by a collaborative approach. Students often have difficulties outside of class. Openings can lead to a discussion of these problems by the teacher and student in a nonthreatening way and additional support from other student services units in such areas can be a beneficial by-product.

KEYWORDS

- **collaboration**
- **digitization**
- **learning**
- **methodology**
- **multimedia**
- **teaching**

REFERENCES

1. Bruner, J., (1985). Vygotsky: An historical and conceptual perspective. *Culture, Communication, and Cognition: Vygotskian Perspectives* (pp. 21–34). London: Cambridge University Press.
2. Cohen, B. P., & Cohen, E. G., (1991). In: Lawler, E. J., (ed.), *From Group Work Among Children to R&D Teams: Interdependence, Interaction and Productivity.*

3. Astin, A. W., (1977). *Four Critical Years: Effects of College Beliefs, Attitudes, and Knowledge*. San Francisco, USA. Jossey Bass Publishing.

4. Pannitz, T. *Collaborative versus Cooperative Learning: A Comparison of the Two Concepts Which Will Help Us Understand the Underlying Nature of Interactive Learning,* 8. https://eric.ed.gov/?id=ED448443 (accessed on 3 November 2020).

5. Gokhale, A. A., (1995). Collaborative learning enhances critical thinking. *Journal of Technology Education, 7,* 22–30. http://scholar.lib.vt.edu/ejournals/JTE/v7n1/gokhale.jte-v7n1.html (accessed on 3 November 2020).

6. Austin, J. E., (2000). Principles for partnership. *Journal of Leader to Leader, 18*(Fall), 44–50.

7. Gokhale, A. A., (1995). Collaborative learning enhances critical thinking. *Journal of Technology Education, 7,* 22–30. http://scholar.lib.vt.edu/ejournals/JTE/v7n1/gokhale.jte-v7n1.html (accessed on 3 November 2020).

8. Bonoma, J., Tedeschi, J., & Helm, B., (1974). Some effects of target cooperation and reciprocated promises on conflict resolution. *Journal of Sociometry, 37*(2), 251–261.

9. Markovsky, B., Ridgeway, C., & Walker, H. *Advances in Group Processes* (pp. 205–226). Greenwich, Connecticut; USA. JAI Publishing.

10. Bean, J., (1996). *Engaging Ideas, the Professor's Guide to Integrating Writing, Critical Thinking, and Active Learning in the Classroom*. San Francisco; USA. Jossey Bass Publishing, USA.

11. Cooper, J., Prescott, S., Cook, L., Smith, L., Mueck, R., & Cuseo, J., (1984). *Cooperative Learning and College Instruction-Effective Use of Student Learning Teams* (pp. 41–65). Long Beach, California; USA. California State University Foundation Publishing, USA.

12. Annett, N., (1997). *Collaborative Learning: Definitions, Benefits, Applications, and Dangers in the Writing Center*. The University of Richmond, Virginia.

CHAPTER 14

A Survey on Forecasting of Currency Exchange Rate Methodologies

MANASWINEE MADHUMITA PANDA,[1] SURYA NARAYAN PANDA,[1] and PRASANT KUMAR PATTNAIK[2]

[1]Department of Computer Science and Engineering, Chitkara University Institute of Engineering and Technology, Chitkara University, Punjab, India, E-mail: mmpandacet@gmail.com (M. M. Panda)

[2]Department of Computer Science and Engineering, KIIT University, Bhubaneswar, Odisha, India

ABSTRACT

The currency exchange market is the leading monetary market within the world. In the international monetary market, rate prediction is a vital issue due to the speedy dynamic data changes and an enormous amount of data availability. According to a literature survey so many techniques are used for currency exchange rate prediction. Some of them are either as good as random walk forecasting models or barely worse. A few researchers said that this displays the proficiency of the exchange market. Different new approaches for prediction exchange rate from 1998 to 2019 are taken into consideration.

14.1 INTRODUCTION

Currency exchange means conversion of one country currency against other. The facts related to economic time series with noisy, volatile, fluctuating, change charge forecasting is a key trouble that has been considered through scholars and experts drastically. Currency conversion is affect by means of a spread of things which includes political and monetary events, or even the psychological country of individual investor and investors. It is not shock that the financial market has taken so much attention for analysis, prediction of future values and trends.

Traditional approaches are not performing well in this field. All suggest that due to unavailability of complete information their approaches will not perform successfully. The successful prediction of currency conversion is based on historical data. One historical time series data assimilates all essential features and these features play an important role for successful forecasting. As money play a vital role in our life so, it is very important to use a suitable tool or best method for currency exchange rate prediction. ANN plays an important role in forecasting currency exchange rates.

To produce a powerful device for time collection prediction using neural network methods in present days, multi-variable nonlinear structures used. Various forecasting methods have been suggested and implemented. Caginal and Laurent formulate statistical validity of any price pattern which has known as the first scientific test pattern [1]. As the data related with financial time series are strident, unbalanced, unstable Giles, Lawrence, Tasoi provide a new intelligent system which reports the difficulties and forecast the foreign exchange rates [2].

14.2   TECHNIQUES USED FOR EXCHANGE RATE PREDICTION

Over the last few years, a large number of methods have suggested and implemented to analyze and forecast financial market movement. A detail discussion about the exchange rate prediction techniques provided here. Zhang, Coggins, Jabri, Dersch, Flower evaluate the forecasting performance of a financial time series uses as a realistic money management system and trading model after Combining wavelet techniques and neural networks [3]. Kamruzzaman, Sarker used artificial neural network (ANN) for forecast conversation rate prediction using three algorithms: BPR (Back propagation with Bayesian Regularization), SBP (standard back propagation), and SCG (scaled conjugate gradient) from three SCG based model outperforms other models [4]. Price prediction is a very difficult task. So to predict the stock price data another technique is used. Shastri, Roy, Mittal proposed a stock market data prediction model. In which the sentiment scores are first calculating through Naive Bayes classifier, then both sentiment scores and historical stock data taken as input on neural network model. In this model, two different scenarios have been taken, one with training on longer period of data (three years data) and another on shorter period of data (one year data). Therefore, after experiment the accuracy of first case is 91% and in second case the accuracy is 98%. In the proposed prediction model Stock market data's pre-processing with the help of HIVE, Hadoop ecosystem [5]. Ince, Trafal used some parametric and non-parametric methods for exchange rate forecast. The parametric methods are: ARIMA, VAR, and co-integration techniques; and the non-parametric methods are: ANN, SVR. ARIMA known as "autoregressive integrated moving average," VAR stands for "Vector Autoregressive," ANN known as "artificial neural network" and SVR stands for "support vector regression." From all of the above techniques SVR technique performance better than other [6]. Subsequently, Rafal, Baikunth, and Michel implement a blending model by joining hidden Markov model (HMM), ANN, and genetic algorithms (GA) to estimate economic market performance. The model shows the price difference between matched days and next day. However, this may not be suitable for some instances [7]. Hanias and Curties use chaos theory for time series prediction. It is used for short-term prediction [8]. Liu Zheng forecast Chinese Yuan (CNY) currency rate using RBF neural network. They can predict CNY spot exchange rate using daily closing prices [9]. Patra, Thanh, Meheruse FLANN (functional link artificial neural network) model for forecast the next day stock price. FLANN shows better result as compared with MLP [10]. Naeini, Taremian, and Hashemi use an Elman recurrent network, a feed forward multi-layer perceptron (MLP) for forecasting a company's stock price [13]. Pacelli, Bevilacqua, Azzollini predict European euro to US dollar up to three days ahead prediction using ANN model [14]. Pacelli provides a comparative analysis of various models like ANN, ARCH, and GARCH used to forecast daily exchange rate from Euro/USD. Among three models ARCH is a good model for prediction [15]. Georgios, Vasilakis, Konstantinos. Theofilatos ·Efstratios, Georgopoulos, Andreas Spiros Likothanassis applied genetic programming approach when forecast EUR/USD in trading. They forecast the next day return when trading [19]. Reheman, Khan, Mahmud implement another method for forex prediction known as recurrent Cartesian genetic programming evolved artificial neural network (RCGPANN). They predict a five-currency rate against the Australian dollar [20]. S. Kumar Chandar, M. Sumathi, S. N. Sivanandam they apply three different learning algorithm for exchange rates between

INR (Indian Rupee) and four other main exchanges Euro, pound sterling (PS), Japanese Yen, US Dollar. It is observed that among all algorithms resilient back (RB) propagation prediction provide more accurate result [21]. Bal, Demir and Aladag used feed forward neural network (FFNN) for Forecast EURO/USD exchange rate and provide different selection criteria to choose different model [24]. N. D'Lima, S. S. Khan proposed forecast model is the combination of an ANN and a hybrid adaptive neuro-fuzzy inference system (ANFIS). This model is used to estimate the future amount of the FOREX market [25]. K. V. Bhanumurthy provides a new approach using a Feed Forward Back Propagation neural network with gradient descent using Levenberg-Marquardt Algorithm which forecasts the price of the USD/INR conversation rate. The performance of the model measures using three evaluation conditions Mean Square Error, Correlation Coefficient, and mean absolute error (MAE) [26]. S. Galeshchuk predicts daily, monthly, and quarterly exchange rates using neural network of three different currencies against USD. The multicurrency's consider for conversation rate are JPY (Yen), EUR (Euro) and GBP (Pound). According to results it display that the short-term forecast procedure provides better exactness in the currency exchange prediction [27]. Bal and Demir shows a comparative evaluation of various activations functions, learning algorithms used in neural network for multicurrency exchange rate. The outcomes determine the ideal parameter combination for ANN modeling of the conversion rate [28]. D. K. Sharma, H. S. Hota, R. Handa apply regression and ensemble regression technique for foreign exchange rate prediction. Result shows that ensemble regression technique perform better than regression technique [29]. J. Bozic and D. Babić make use of wavelet rework and synthetic neural networks (ANNs) for the forecast of monetary time collection [30].

The main approach/system used for achieving forecast exchange rate prediction system based on neural network building blocks. ANN plays a dynamic role in overseas exchange rate estimate procedure. It is a nonlinear, data driven and nonparametric modeling method. Chandar, Sumathi, Sivanandam had used ANN model with various learning parameter for exchange rate prediction. Using ANN method they are forecasting Indian Rupees (INR) against four other currencies such as Japanese Yen (JYEN), PS, EURO, and United States Dollar (USD) over 100 days. They use five different training algorithms (GDM-gradient descent with momentum, RP-Resilient Back propagation, GDA-Variable

Learning Rate Back-propagation, LM-Levenberg-Marquardt, GD-Batch gradient descent. for exchange rate prediction. To predict the currencies they considered 1205 days data. Among these data they had used 80% for training and 20% for testing. After applying five different algorithms the result shown that Levenberg-Marquardt based model reaches nearer prediction for all currencies than other models [23]. Lin, Chen, Lo use a prediction system based on regression technique for currency conversion. They use an IERPS (Intelligent Exchange Rates Prediction System) system for forecast Taiwan Dollar (TWD) to Pound of British (GBP), Dollar of Australian (AUD), Euro (EUR), and Yen of Japanese (JPY). To do this experiment they had collected data from Jan-02-2006 to Mar-31-2011. After applying this technique they average accuracy is 97.47% [16].

Rout, Majhi, Majhi, and Panda apply different methods of ARMA for yen, pound, and rupees interchange rates with respect to the US dollar. They had use ARMA-PSO (particle swarm optimization), ARMA-CSO (cat swarm optimization), ARMA-BFO (bacterial foraging optimization) and ARMA-FBLMS (forward backward least mean square) and ARMA-differential evolution (DE) for exchange rate prediction. After applying various methods for exchange rates prediction the proposed DE-based adaptive parameter in ARMA model perform best compared to other three similar models [22]. Tlegenova use ARIMA (autoregressive integrated moving average) model for currency conversion of Dollar-to-Kazakhstan (USD/KZT), Euro-to-Kazakhstan (EUR/KZT) and Singapore dollar-to-Kazakhstan (SGD/KZT). To do this experiment they had collected data from 2006 January to 2014 December. The model performance is measured by computing MAE, MAPE, and RMSE value [31].

Xhaja, Ktona, Brahushi use support vector machine, and ANN for currency conversion of EUR/ALL (Euro-to-All). To perform this experiment data collected from January 2010 to December 2013. It shows that SMOReg (SVM for regression) perform better than ANN [5]. Hua, Zhang, Leung used FLANN based on kernel regression technique for currency conversation rate prediction. The exchange rate US dollar against Indian Rupees and Japanese Yen, British Pound. They had used two methods (FLANN without KR and FLANN with KR) for exchange rate prediction. FLANN-KR Models performs better than FLANN [12].

Majhi, Panda, Sahoo use FLANN and CFLANN to predict exchange amount between US$ to JPY (Japanese Yen), British Pound and INR (Indian Rupees). To evaluate the performance they had taken data from

January 1973 to January 2005 for dollar to rupees, January 1971 to January 2005 for dollar to pound, January 1971 to January 2005 for dollar to Japanese Yen [11]. According to prediction result CFLANN perform better than FLANN model. Nagpur proposed deep learning model using SVR, ANN, and long short-term memory (LSTM) Neural Network with Hidden Layers for exchange rate of multicurrency against US Dollar. He considered EUR, CNY, AUD, JPY, CHF, GBP, INR, SEK, MXN, NZD, and CAD [32].

14.3 FREQUENTLY USED MODELS FOR EXCHANGE RATE PREDICTION

For currency conversation rate prediction usually ANN, autoregressive moving average (ARIMA), functional link ANN, and deep neural network (DNN) models are used.

14.3.1 ARTIFICIAL NEURAL NETWORKS (ANNS)

When soothing the linear constraints of the model, the possible number of nonlinear structures that can be used to describe and predict the time series will be very large. A good quality nonlinear model should be sufficient to capture a few of the nonlinear phenomena in the data. ANN is one such model that can approximate a variety of nonlinearities in the data [28].

ANN is a deep learning scheme that works in the same fashion as that of human brain. In this, the units are interconnected in the similar way as that of artificial neurons given by Hinton [37]. The neurons in ANN are mainly comprises of three main units such as: synapses utilized for connecting units, a summer which added input data weighted by synapses and an activation function employed for transforming the desired data at the output layer as depicted in Figure 14.1.

Mathematically, ANN can be represented as:

$$V_k = \sum_{j=1}^{n} W_{kj} y_j$$

$$Z_k = \varphi(V_k + c_k)$$

where;

$y_i = y_1, y_2, y_3, \ldots\ldots\ldots\ldots\ldots\ldots y_n$ are the input signal;

$W_{kj} = W_{k1}, W_{k2}, W_{k3}$ .... W_{kn} are the weight of synaptic;
V_k signifies the resultant output of the summer;
c_k denotes the bias;
φ denotes the activation function; and
Z_k neuron's output signal.

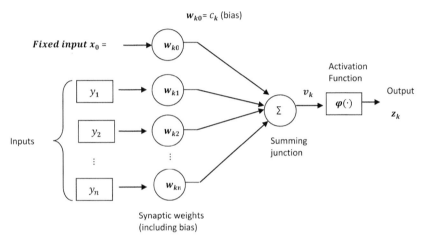

FIGURE 14.1 ANN mathematical model [35].

The neurons with the same properties are grouped into a single layer. The layer is the representation of set of nodes that are used to forward information to the other linked layer as well as the external environment. The neural network with single layer represents the simplest and the easiest form of ANN structure. The first and the last layer consist of the input nodes and the output nodes, respectively. Based on the input and the output parameters the amount of input and output layer nodes decide. For the output layer the input data will provided after the input data is processed in the hidden layer.

14.3.2 AUTO REGRESSIVE INTEGRATED MOVING AVERAGE (ARIMA) MODEL

ARIMA is a clean stochastic time series version that we are able to use to train and then forecast future time points. An ARIMA model is a class of statistical approaches for analyzing and estimating time collection information. It clearly provides a collection of preferred systems in time series statistics, and as such gives a simple yet effective approach for creating expert time series forecasts.

This abbreviation is descriptive, taking the important features of the model itself. In detail, they are:

1. **ARIMA is Auto Regressive (AR):** A version that makes use of the dependent association between an observation and some variety of lagged observations. **P** represent the number of lagged observations integrated in the model.

2. **Integrated (I):** The use of differencing of raw observations (e.g., subtracting a statement from an observation on the preceding time step) which will make the time series stationary. The degree of differencing represented by D.

3. **Moving Average (MA):** The dependency present between an observation and a residual error from a MA model applied to lag observation through a model. Q represents the order of MA or also known as the size of MA.

   The three properties of ARIMA model are P, D, and Q.

14.3.3   FUNCTIONAL LINK ARTIFICIAL NEURAL NETWORK (FLANN) MODEL

To overcome the difficulties connected with multi-layer neural network, single layer neural network are often taken into thought as an opportunity technique. Due to the single layer neural network linearity in nature so; it fails to map the complicated nonlinear issues. It is a big challenge to resolve such issues using feed forward ANN having single layer. The FLANN model is suggested to maintain the hole between computation intensive and complex multi-layer neural network and the linearity of the single layer neural network. A single layered non-linear neural network is known as FLANN. The hidden layer is eliminated in FLANNs without giving up non-linearity by providing the enter layer with expanded inputs which can be built as the capabilities of unique attributes. Elimination of hidden layer makes those networks extraordinarily easy and simple reasonably priced. It performs better than MLANN. Nonlinear growth will make. The development effectively will build the dimensionality of the info vector and henceforth, the hyperactive planes created by the method for the FLANN give more prominent segregation ability in the input pattern zone.

The FLANN structure is depicted in Figure 14.2. Here, the practical extension square makes utilization of a utilitarian model including a subset of symmetrical sin and cos premise capacities and the first example alongside its external items. For instance, seeing a two-dimensional info design X = [x1, x2]T. The improved example is acquired by utilizing the trigonometric capacities as X = [x1cos(πx1) sin(πx1).... x2cos(πx2) sin(πx2)... ... x1x2] T, which is then utilized by the system for the adjustment reason.

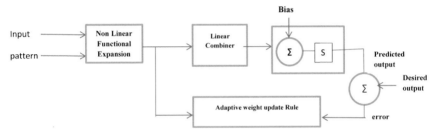

FIGURE 14.2   Block diagram of FLANN.

After enlargement the number of input elements is more than the original input. The enter factors are extended with weight and the sum is calculated. Every extended inputs are weighted, then summed together with the weighted bias and finally, its miles passed through tanh (.) nonlinear characteristic for offering the expected productivity.

In designing the adaptive predicting model, it's far expected that ok beyond economic data is without a problem accessible for every day. From those records applicable abilities similar to each day are extracted and arranged. For a selected monetary estimating case permit Z to show the number of enter sample from which R used for number of training patterns and T used for no of testing patterns of the recommended model.

Thus Z=R+T. Let X be the n-dimensional input vector given:

$$X=[x_1 \ x_2 ... x_n] \qquad (1)$$

Each component xi (i=1, 2,..., n) the enter vector will extended into 2k+1 term.

$$x_i'=[x_i \sin(\pi x_i)\cos(\pi x_i)\sin(3\pi x_i)\cos(3\pi x_i)\text{--------}\sin(k\pi x_i)\cos(k\pi x_i)] \quad (2)$$

The output of the extension block with n(2k+1) element is ŷ(n).

$$\hat{y}(n) =[x_1,' x_2'...x_n']$$ (3)

Let W be the weight vector with n(2k+1) elements and according to the unit bias w_b be the weight shown in Figure 14.3.

$$W = [w_1, w_2...w_{n(2k + 1)-1,}\ W_{n(2k + 1)}]$$ (4)

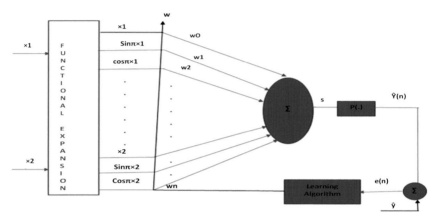

FIGURE 14.3   FLANN structure.

The output of the model ŷ(n) produces according to input and it acts as an evaluation of the desired price. The linear part of the model's productivity is calculated as:

$$S = W.x_n+w_b$$ (5)

where, w_b represents the weighted bias (may be positive or negative). This output is then surpassed through a nonlinear characteristic (a sigmoid characteristic) to yield the expected output. The difference between the model output and the desired response is represented as error sign e(n).

$$e(n) = \hat{y}(n)-y(n)$$ (6)

To compute the correction weight vector W weight update algorithm is used. The input vector X and error e(n) are employed for that.

$$W = \mu.X.e(n)$$ (7)

The learning coefficient, μ wheels the convergence rate and lies between 0 and 1.

14.3.4 DEEP NEURAL NETWORK (DNN)

Deep learning is a field of study in machine learning established on a class of statistical learning algorithms loosely modeled after the neurons in the human brain which use multiple levels of nonlinear operations to find patterns in complex data such as images for example. Then non-linear operations are a crucial aspect that allows neural networks to model extremely complex structures. Previous to this discovery, it was considered too difficult to train neural networks with many layers due to a problem known as the "vanishing gradient problem." Deep learning algorithms present data over a hierarchical learning method after remove high level and complex ideas [33]. A neural network is measured to be "deep" as long as it has more no of hidden layer.

Figure 14.4 depicting the different layers of DNN. The bottom FFNN is not "deep" since it only has one hidden layer. However, the top FFNN is "deep" since it has many hidden layers. Since this discovery, a lot of breakthroughs have been made within deep learning in almost every aspect. In the new millennium, DNNs have ultimately attracted extensive-spread attention, especially through the usage of outperforming opportunity machine mastering strategies [34]. It's regarding advances in neural network architectures, different methods for training neural networks or discovering new applications for deep learning in the real world. Because the reason deep learning has become as popular as it is today is due to its successful applications in fields such as computer vision, handwriting recognition, natural language understanding, speech recognition, audio processing, robotics, information retrieval and more. Deep learning has also been used successfully to win many competitions within topics such as visual mitosis detection and optical character identification.

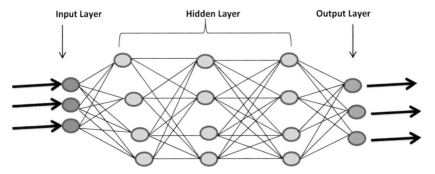

FIGURE 14.4   Deep neural network [36].

14.4 EVALUATION CRITERIA

To estimate the forecasting performance MAE, and the mean absolute percentage error (MAPE), root mean square error (RMSE) values calculated.

MAE measures the average degree of the errors in a set of predictions. It's the average of the total differences between expectation and actual observation where all individual differences have equivalent weight.

$$\text{MAE} = \frac{1}{N} \sum_{t=1}^{n} \left| y_t - \hat{y}_t \right| \tag{8}$$

where, N: number of input; y_t and $\hat{y}_t$ are the actual value and prediction value.

RMSE It will use to check the performance of the model. It does this by measuring difference between expected values and the actual values.

$$\text{RMSE} = \sqrt{\frac{1}{N} \sum_{t=1}^{N} (y_t - \hat{y}_t)} \tag{9}$$

where, N: number of inputs; y_t and $\hat{y}_t$ are the actual value and prediction value.

MAPE is the totality of the single absolute errors divided by the demand (each epoch separately). Actually, through this the average of the percentage errors will calculated:

$$\text{MAPE} = \frac{1}{N} \sum_{t=1}^{N} \left| \frac{y_t - \hat{y}_t}{y_t} \right| \times 100\% \tag{10}$$

where, N: number of inputs; y_t and $\hat{y}_t$ are the actual value and prediction value.

14.5 CONCLUSION

In this study we have given a wide survey of the most important article published in the literature from 1998–2019. The review based on prominent work done on prediction exchange rate. For each of the considered articles the existing problems that to be resolved were discussed. The review focused on different methods used for forecast exchange rate prediction and different parameters to evaluate the performance of different model. Using these parameters we can choose the best method for prediction of exchange rate. Based on this compressive review, the authors believe that currency exchange prediction play a vital role in monetary market. Therefore, it aims to be a major development in exchange rate prediction.

KEYWORDS

- **artificial neural network (ANN)**
- **deep neural network (DNN)**
- **exchange rate**
- **genetic algorithms**
- **multilayer perceptron**
- **resilient back**

REFERENCES

1. Caginal, P. G., & Laurent, H., (1998). The predictive power of price patterns. *Applied Mathematical Finance, 5,* 181–205.
2. Giles, C. L., Lawrence, S., & Tasoi, A. C., (2001). Noisy time series prediction using a recurrent neural network and grammatical inference. *Machine Learning, 44*, 161–183.
3. Zhang, B. L., Coggins, R., Jabri, M. A., Dersch, D., & Flower, B., (2001). Multiresolution forecasting for futures trading using wavelet decompositions. *IEEE Transactions on Neural Networks*, 765–775.
4. Kamruzzaman, J., & Sarker, R. A., (2004). *ANN-Based Forecasting of Foreign Currency Exchange Rates, 3*(2), 49–58.
5. Shastri, M., Roy, S., & Mittal, M., (2019). Stock price prediction using artificial neural model: An application of big data. *EAI Endorsed Transactions on Scalable Information Systems, 1*, 1–8.
6. Ince, H., & Trafalis, T. B., (2006). *A Hybrid Model for Exchange Rate Prediction* (pp. 1054–1062). Elsevier.
7. Hassan, M. R., Nath, B., & Kirley, M., (2007). *A Fusion Model of HMM, ANN and GA for Stock Market Forecasting* (pp. 171–180). Science Direct.
8. Mike, P. H., & Panayiotis, G. C., (2008). Time series prediction of dollar\euro exchange rate index. *International Research Journal of Finance and Economics.* pp. 232–239.
9. Liu, Z., Zheng, Z., Liu, X., & Wang, G., (2009). *Modeling and Prediction of the CNY Exchange Rate Using RBF Neural Network* (pp. 38–41). IEEE Transaction.
10. Patra, J. C., Thanh, N. C., & Meher, P. K., (2009). Computationally efficient FLANN-based intelligent stock price prediction system. *International Joint Conference on Neural Networks* (pp. 2431–2438). Atlanta, Georgia, USA, IEEE.

11. Majhi, R., Panda, G., & Sahoo, G., (2009). *Efficient Prediction of Exchange Rates with Low Complexity Artificial Neural Network Models* (pp. 181–189). Elsevier.

12. Hua, X., Zhang, D., & Leung, S. C. H., (2010). Exchange rate prediction through Ann based on kernel regression. *Third International Conference IEEE* (pp. 39–43).

13. Naeini, M. P., Taremian, H., & Hashemi, H. B., (2010). Stock market value prediction using neural networks. In: *Computer Information Systems and Industrial Management Applications (CISIM), 2010 Int. Conf.* (pp. 132–136).

14. Pacellil, V., Bevilacqua, V., & Azzollini, M., (2011). Scientific research. *Journal of Intelligent Learning Systems and Applications,* (pp. 57–69).

15. Pacelli, V., (2012). Forecasting exchange rates: A comparative analysis. *International Journal of Business and Social Science, 3*(10).

16. Lin, S. Y., Chen, C. H., & Lo, C. C., (2013). Currency exchange rates prediction based on linear regression analysis using cloud computing. *International Journal of Grid and Distributed Computing*, 6.

17. Onder, E., Bayır, F., & Hepşen, A., (2013). Forecasting macroeconomic variables using artificial neural network and traditional smoothing techniques. *Journal of Applied Finance and Banking, 3*(4), 73–104.

18. Stepnika, M., Cortez, P., Donate, J. P., & Stepnika, L., (2013). *Forecasting Seasonal Time Series with Computational Intelligence: On Recent Methods and the Potential of Their Combinations* (pp. 1981–1992). Elsevier.

19. Vasilakis, G. A., Theofilatos, K. A., Georgopoulos, E. F., Karathanasopoulos, A., & Likothanassis, S. D., (2013). *A Genetic Programming Approach for EUR/USD Exchange Rate Forecasting and Trading* (pp. 415–431). Springer Science.

20. Reheman, M.,Khan, G. M., & Mahmud, S. A., (2014). *Foreign Currency Exchange Rates Prediction Using CGP and Recurrent Neural Network* (pp. 239–244). Elsevier.

21. Kumar, C. S., Dr. Sumathi, M., & Dr Sivanandam, S. N., (2014). Neural network-based forecasting of foreign currency exchange rates. *International Journal on Computer Science and Engineering (IJCSE), 6*(6).

22. Rout, M., Majhi, B., Majhi, R., & Panda, G., (2014). Forecasting of currency exchange rates using an adaptive ARMA model with differential evolution-based training. *Journal of King Saud University Computer and Information Sciences*, pp. 7–18.

23. Dr. Chandar, S. K., Dr. Sumathi, M., & Dr Sivanandam, S. N., (2015). *Forecasting of Foreign Currency Exchange Rate Using Neural Network, 7*(1).

24. Bal, C., Demir, S., & Aladag, C. H., (2016). A comparison of different model selection criteria for forecasting EURO/USD exchange rates by feed forward neural network. *Int'l. Journal of Computing, Communications and Instrumentation Engg. (IJCCIE)*, 3.

25. D'Lima, N., & Khan, S. S., (2016). *FOREX Rate Prediction using ANN and ANFIS.* IEEE.

26. Bhanu, M. K. V., (2016). Forecasting foreign exchange rate during crisis: A neural network approach. *International Proceedings of Economics Development and Research IPEDR* (Vol. 86).

27. Galeshchuk, S., (2016). *Neural Networks Performance in Exchange Rate Prediction* (pp. 446–452). Elsevier.

28. Bal, C., & Demir, S., (2017). A comparative study of artificial neural network models for forecasting USD/EUR-GBP-JPY-NOK exchange rates. *Proceedings of the*

Second American Academic Research Conference on Global Business, Economics, Finance and Social Sciences (pp. 28–30). New York-USA.

29. Sharma, D. K., Hota, H. S., & Handa, R., (2017). Prediction of foreign exchange rate using regression techniques. *Review of Business and Technology Research, 14*(1).

30. Bozic, J., & Babić, D., (2018). Financial time series forecasting using hybrid wavelet-neural model. *The International Arab Journal of Information Technology, 5*(1).

31. Tlegenova, D., (2019). *Forecasting Exchange Rates Using Time Series Analysis: The Sample of the Currency of Kazakhstan, arXiv, 1–8.*

32. Nagpure, A. R., (2019). Prediction of multi-currency exchange rates using deep learning. *International Journal of Innovative Technology and Exploring Engineering (IJITEE), 8*, 1–7.

33. Najafabadi, M., Villanustre, F., Khoshgoftaar, T., Seliya, N., & Muharemagic, R., (2015). Deep learning applications and challenges in big data analytics. *Journal of Big Data 2*(1), 1–21.

34. Schmidhuber, J., (2014). Deep learning in neural networks: An overview. *Neural Networks, 61*, 85–117.

35. Zhang, G. P., (2003). Time series forecasting using a hybrid ARIMA and neural network model. *Neurocomputing, 50*, 159–175.

36. https://alphabold.com/neural-networks-and-deep-learning-an-overview/ (accessed on 3 November 2020).

37. Hinton, G. (1982). *How Neural Network learn from Experience,* Scientific American, pp. 145–151.

e-Governance, Issues, and Challenges of m-Governance in India

SARITA DHAL, DEEPANJALI MISHRA, and NISHIKANTA MISHRA

School of Humanities, Kalinga Institute of Industrial Technology, Deemed to be University, Bhubaneswar, Odisha, India

ABSTRACT

m-Governance is the modern information and communication technology for reaching the unreached. It is a system which utilizes the Internet, local area network (LAN), and mobile device for the purpose of improving effectiveness, efficiency, and delivery of services to the common people. It is the governance through electronic media to promote and justify democracy. In a democracy, the primary job of the government is to focus on society on achieving the public interest. Governance is a means of the link between government and political, social, and administrative environment. e-Governance can transform citizen service in order to empower them. It helps the public to participate in government and enhance citizens' economic and social opportunities. This will help the society to lead a better life not only for the present period but also for the future. m-Governance otherwise known as governance through mobile devices/ phones is a domain of e-Governance. It is more popular due to its low cost and easy accessibility. It is the cheapest way of communication even in rural areas [1]. It can provide timely and accurate information which can lead to public empowerment. m-Governance is one web approach of the government in all the departments by using mobile device. Mobile seva delivery gateway can be one of to its cost saving, proficiency, convenient, and flexibility, easy, and better access even in remote rural areas. Different factors like rules and regulations, establishment of information security system, optimization of business process, and evaluation of e-Governance are responsible in order to have a reliable and good m-Governance [2].

15.1 INTRODUCTION

e-Governance is one of the easy and costs less information services provided by the government to the common people. m-Governance is a sub-domain of e-Governance. It ensures that electronics services are available to the common man by using mobile phones to serve quickly and to by-pass the traditional network of communication. In India, m-Governance plays an important role and more popular in all government department due to its easy accessibility. At the same time mobile phone service is the cheapest way of communication in rural areas of the country.

Now the government is promoting to deliver e-Governance services through mobile phones. In India mobile phones has become a revolution even in remote rural areas where the basic facility like electricity connection is not proper. m-Governance service is more popular in India due to increase in accessibility of mobile phones, millions of subscriber base and easy adoptability.

15.2 OBJECTIVES OF THE STUDY

The present study is based on the following objectives:

1. To highlight the motives of the government behind use of m-Governance service.
2. To find out the measures taken by the government for m-Governance system.
3. To know the merits of m-Governance service.
4. To highlight the challenges faced by the government for a reliable m-Governance.
5. To develop different determining factors for good m-Governance.

15.3 METHODOLOGY

The research methodology of the study is based on purely secondary data. The data are collected from journals, magazines, and research articles available in the website. Some of the data in this study are collected from the web site of the ministry of electronics and IT, Government of India.

15.4   WHY M-GOVERNANCE?

Facilitates the banking sector for financial transactions. The m-Governance service is used by the government due to the following grounds. The objectives of m-Governance service are:

1.   Provide fast and cost less access of public services.
2.   Provide timely and accurate information of the government.
3.   Helps empowering common people in the country.
4.   Prominently used by all the Government Departments of the country.

15.5   AS AN EMERGING SERVICE CHANNEL

Nowadays, mobile phones are not only limited with a tool for communication but also has been emerged as a strongest technology to bridge the digital divide between the have and have-not, urban, and rural, literate, and illiterate, men, and women. Within a very short span of time its introduction has make a significant development even in remote rural areas where difficulties like lack of electricity and power, lack of fully fledged connectivity as well as lack of low level literacy rate. At the same time it has created job opportunity either directly or indirectly to the rural youth.

m-Governance is not restricted to the above fact now it has emerged as a delivery channel for all type of government and private service. Now a mobile holder can transfer his money from one account to another account of different banks. The business house use mobile services to its customers and traders through mobile devices. The Reserve Bank of India (RBI) has also permitted to all commercial banks of the country to provide various banking services thorough mobile phones [4].

The present central government has formed an inter ministerial department which is responsible to deliver all financial services to the common people through mobile phone. The ministry of communication of India has launched a mobile Seva Abhiyan with an aim to provide different government services to the general public of the country. Accordingly infrastructure development has been made in the country in order to enable the people to get various welfare services through mobile devices. As a result 3G technologies have been developed in India. The purpose of this technology is to enable the people to access easily the government services like health, education, agriculture, finance, and environment as well as whether information through mobile phones.

An inter-ministerial department has been created between the ministry of communication and IT and the ministry of electronics and IT by the Indian government. This department has developed a strategy to provide round the clock access to the public by mobile phone throughout the country [5]. An unique infrastructure and application development ecosystem has been created for m-Governance in the country. As a result the rural people can able to easily access the public services through wireless technology.

15.6   MEASURES TAKEN BY THE GOVERNMENT

The following are some of the measures taken by the government in order to promote m-Governance:

1. All the government department and their agency have to create a website which should be made mobile complaint using one web-approach.
2. Mobile application can be ensured access various operating agencies as per the government policy on open standards of e-Governance.
3. For the convenient use of mobile phones, uniform single pre-designated numbers having long and short code shall be used for mobile-based services.
4. All government and different agencies under the department should develop mobile application for all their services through mobile devices. At the same time the department must have to specify the service level for such services so that the general public can easily access different government services.
5. To ensure and adoption of this policy in a time-bound manner the government has to develop a mobile seva delivery gateway (MSDG) which will ensure the availability of public services through mobile devices.

15.7   CONCEPT OF MOBILE SEVA DELIVERY GATEWAY (MSDG)

This is a technical infrastructure for integration of mobile application with common e-Governance infrastructure to delivery public services to the users of mobile phones. It will help the government and its agencies to implement e-Governance services easily. At the same time it will act as a cost-saving device for m-Governance services thorough out the country. This platform can enable the government to fulfill the demand for e-Governance services [6]. It can easily maintain m-Governance services though out the country.

MSDG shall develop and deploy mobile-based application for government services. As mobile-based technology is growing rapidly, more channels may be used in future due to the following requirements:

1. SMS (short message services);
2. IVR (interactive voice response);
3. WAP (wireless application protocol);
4. USSD (unstructured supplementary service data);
5. CBC (cell broad cost);
6. Others like: WIFI, WLAN, etc.

15.8   MERITS OF M-GOVERNANCE

Some of the advantages of m-Governance are mentioned below:

1. **Cost Saving:** m-Governance is one of the low cost services provided by the government to the general public. As there are lots of mobile network available there is a competition between the mobile companies to use their network respectively. As a result transfer of message can be possible with a low cost.
2. **Proficiency:** m-Governance can enable the public to know the proficiency of different development programs of the government. As transfer of various welfare schemes meant for the public can be made through mobile phones its proficiency can be realized by the general public even in the remote areas of the country.
3. **Convenience and Flexibility:** It is considered that m-Governance system is one of the most convenient methods for transferring message towards welfare program of the government in different times. It is a more transparent method of communication for

the implementation of these programs. The public can easily be communicated the contents of the welfare scheme and the implementation process.

4. **Better Service:** It has been considered that there is no other better service available with the government except m-Governance system. This system is more convenient which can provide better service to the general public of the country in a short span of time.

5. **Easy Access:** Today, it can be said that there is a mobile revolution in the country. Access to mobile phone is not only limited with the citizens of urban areas but also spread to even rural areas of the country as well. It is possible due to its easy accessibility. As a result different mobile networks are connected even in remote rural areas of the country.

6. **Quick Interaction:** m-Governance system can enable quick and easy interaction between the government and the general public. As a result the welfare programs of the government can be interacted with the public directly for which the government can be accountable in near future [7]. This quick interaction resulted in awaking the mass of the country towards different public welfare programs.

De-Merits of m-Governance

The following are some of de-merits of m-Governance system:

1. **Complexity of Different Mobile Technology:** This is considered one of the major drawbacks in the system. Different mobile network has their own technology. Hence it seems to be difficult on the part of a particular department to adopt all these technology for the public. This complexity of the technology create problem for transfer of message through mobile phones.

2. **Creation of Secured Network:** Creation of a secured network to deliver reliable services is a major problem for the government. As different mobile phone with different network are used by the public both in urban and rural areas it seems to be difficult on the part of the government as well as all its agencies to find put a secured and reliable network which can be utilized for transmission of public message.

3. **Identification of Easy Service Provider:** This is one of the major demerits with m-Governance system. The entire government

department as well as its agencies cannot able to find out a uniform service provider in order to transfer the message to public uniformly. Different parts of the country cannot be communicated with different type of information in relating to the public service program.

4. **Chances of Legal Issues:** m-Governance is meant to serve the general public of the country in a better manner. However, if anything goes wrong either intentionally or unintentionally the matter will be sub juiced in the court of law. Again with the introduction of right to information (RTI) the message can be brought to legal compliance.

5. **Chances of Password Cracking and Virus:** This can be another possibility of drawback in m-Governance system. Due to misuse of different web site it can be possible on the part of the hackers to crack the password created by different departments of the government. Misuse of web technology the virus in the web site will be a regular and continuous problem for the government.

15.9 FACTORS TO IMPROVE M-GOVERNANCE

The government is considered m-Governance is one of the easiest and quick methods of transferring all welfare programs meant for the public. Hence a regular effort has been made by the government and its agency how to improve this system in the country [8]. The following are some of the determining factor for an improved m-Governance system:

1. **Uniform Rules and Regulation:** The government and its agency must have to prepare a uniform rules and regulation for an effective m-Governance system in the country. A uniform standard should be prepared by all government department so that the drawbacks of m-Governance system can either minimized or eradicated while transferring the message to the general public.

2. **Information Security System:** The information security system should be strengthened in the country for a better m-Governance. A reliable information system should be created so that the confidential information should not be transferred in the name of m-Governance. On the other hand all the public utility services of the government should not be misused and propagated through

m-Governance which can misled the general public. Hence it is desirable to strengthen the information security system of the country for a reliable and dependable m-Governance system.

3. **Optimization of Administrative Business Process:** So far business process is concerned it should be updated. The latest technology in marketing as well as advertisement should be introduced in business administration of the country. Different updated customer satisfaction formula and sales promotion technique should be introduced in the marketing of business policy. Different product policy, price policy, distribution channel, marketing research policy should be updated for better business information through m-Governance.

4. **Strengthening the Evaluation of e-Governance:** Timely evaluation of e-Governance system should be made properly. It ensures that the public utility services are timely and properly accessed by the general public. If necessary some changes should be made in the system so that its benefit can be available to large section of the society.

5. **Infrastructure Facility for m-Governance:** It should be ensured by the government that the infrastructure facility should be upgraded regularly in order to improve the m-Governance service in the country. Due to technological up-gradation in the field of mobile network in the country the government should be careful to update its information transfer technology so that large sections of the society can able to get the benefit of m-Governance service.

15.10 M-GOVERNANCE IN INDIA

Mobile service has emerged as a delivery channel for different e-Governance service in India now. Considering the huge mobile penetration in the country particularly in rural areas, it has become important to offer government services over mobile network devices. Now it is possible to deliver different government services meant for the public at their doorstep [9]. Considering the importance, the ministry of communication and IT with the department of electronics and telecommunication has announced that all its departments and agency have to develop and deploy mobile applications to provide all the related public services through mobile phone.

Considering the increasing number of mobile subscription and its reach it has become essential to offer government services over mobile phones. It ensures that the e-Governance services can be available to each and every citizen at their doorstep.

Mobile phones considered being the largest service access provider for the government services. e-Governance of the country can be exchanged through National Governance Service Delivery Gateway (NSDG) and State e-Governance Service Delivery Gateway (SSDG) both in the central as well as in the state level of the country respectively [10]. Hence m-Governance has become a supplementary service for the Indian government in order to promote e-Governance service in the country. Continuous effort has been made in the central and state government level to improve m-Governance service so that public utility program can reach to the common people of the country in their doorstep. As a result the participatory democracy can prevail in our country, which can eradicate poverty within a targeted period of time.

15.11   CONCLUSION

The following conclusions are derived from the research study:

1. m-Governance service is a domain of e-Governance system of the country in order to transfer public utility service to the common people at their doorstep.
2. It is considered as a costless service provider for all government welfare programs meant for the common people of our country in central and state government level.
3. For proper implementation of m-Governance service in the country the government has undertaken different measures like creation of one-web approach, mobile operating system, and uniform pre-designated numbers and development of MSDG in the country.
4. m-Governance system is very popular in our country due its easy accessibility and convenience in use by the rural poor. Its proficiency and low cost encourage the government to transfer all its public utility services through mobile phone to the common people of the country so that they can participate in the government welfare scheme by promoting democracy.
5. In order to strengthen the m-Governance service in the country the government have to prepare a uniform standard of communication

network throughout the country so that different mobile network can be utilized by the common people easily.

6. Information security act should be strengthening in the country in order make free from hacking and misutilization of confidential messages of the government. Therefore, it can be flexible in operating of the m-Governance system in the country in the future.

7. Business operation system in the country should be updated in order to have an ethical business environment for the benefit of common people, and timely evaluation of e-Governance service should be ensured in both central as well as state government level for improvement of m-Governance service in the country.

8. Mobile-based channels should be upgraded in order to meet the requirements of SMS, IVR, WAP, USSD, CBC, WIFI, WLAN services in the country for effective use of m-Governance service.

9. The government should ensure that the entire government department should develop its technology through the web, which can be available in the mobile network of the country so that the day-to-day development in the department can be made available to the common people of the country.

10. It can be concluded that m-Governance service, if utilized properly, then the participation of common people in the government can increase to a maximum extent, which can solve the purpose of the democratic system of our country. As a result, there will be no doubt that the dream of our father of nation GRAMYA SWARAJ can be fulfilled.

KEYWORDS

- **administrative environment**
- **domain**
- **e-Governance**
- **local area network**
- **m-governance**
- **mobile seva delivery gateway**

REFERENCES

1. Rennu, R., & Mahalakoiv, T., (2010). *Mobile Govt. and Beyond Working Document.* Mobile Solutions Ltd.
2. Budhiraja, R., (2003). *Electronic Governance: A Key Issue in the 21ˢᵗ century.* Electronic Governance Division, Ministry of Information Technology, Government of India. renu@miti.gov.in.
3. Dutton, W. H., Gillett, S. E., McKnight, L. W., & Peltu, M., (2004). Bridging broadband Internet divides: Reconfiguring access to enhance communicative power. *Journal of Information Technology, 19*(1), 28–38.
4. http://egovernancenepal.blogspot.com/2007/04/e-governance (accessed on 3 November 2020).
5. Prensky, M., (2004). What can you learn from a cell phone?-Almost anything. *Journal of Online Education,* 1–9.
6. Ghyasi, A. F., & Kushchu, I., (2004). *m-Government: Cases of Developing Countries.* Mobile Government Lab (mGovLab).
7. El Kiki, T., & Lawrence, E., (2006). Government as a mobile enterprise: Real-time, ubiquitous government. *Proceedings of the Third International Conference on Information Technology: New Generations (ITNG'06).*
8. Kumar, M., & Sinha, O. P., (2007). m-Government-mobile technology for e-government. *International Conference on e-Government* (pp. 294–300). India.
9. Tornatzky, L. G., & Fleischer, M., (1990). *The Processes of Technological Innovation.* Lexington Books, Lexington, Massachusetts.
10. *National e-Governance Plan (NeGP).* Annual Report at: http://www.mit.gov.in/content/national-egovernance-Plan (accessed on 3 November 2020).

CHAPTER 16

'IT' Does Matter: The Smart Solutions

SWAPNAMOYEE PALIT

School of Humanities, Kalinga Institute of Industrial Technology, Deemed to be University, Bhubaneswar, Odisha, India

ABSTRACT

'Bhubaneswar' joined the 'smart city club of 100 cities' only in March 2016. So the impact of this clubbing in its wide bloom can only be known after some more years in terms of its overall influence on total income and job generation as the multiplier process would take its lag. All these developments have the potential to address positively the issues of poverty, unemployment, migration, etc., as it connects the individuals with his social, political, and economic environment from his place of settlement. The G2C (Government to Citizen) and C2C (Citizens to Citizens) connectivity made possible by the wave of technology has further strengthened the democracy with every issue getting overall public vision and opinion. The impact in its wide bloom can only be known after some more years in terms of its overall influence on total income, job generation, economic growth and development, etc., as the multiplier process would take its lag. This chapter seeks to converse about the existing issues and apprehensions related to the smart city and its technological disruption, which confront its economic contributions.

16.1 INTRODUCTION

As the joke on information technology (IT) goes that a person once told his maid that he would pay her online and to this she replied that she would work from home. Reading between the lines are many features of the technological changes brought about by the rapid surge of technology.

Some of these are digitalization of most transactions, easy communication, hassle-free system and a plethora of benefits overshadowing its adverse impact of technological divide between the skilled and the unskilled, the awareness and adaptation to the changed scenario and the rapid displacement of variables factors with automation of every units to speak a few, remaining confined to the taken instance in introduction. So obvious the debate is no more about whether to adopt the technology economy or not because we are already in the surge but rather how to navigate smoothly in this technology economy through optimization of its benefits and positivization of its negative externalities. Here 'positivization' would mean to even out the roughness spilled by the surge of technology through appropriate initiatives at all levels (national, state, company or at individual level) so that everyone gets a share in the benefits. This chapter seeks to converse about the existing issues related to the smart city and technological disruption which sometimes confront its economic contributions.

16.1.1 *REVIEW OF LITERATURE*

Smart city concept relates to the integration of urban informatics of a city to regenerate it with an objective to promote sustainable economic development [6]. Thus it is a relative term which would show variations according to cities and countries depending on factors like their current status of development, adoption of reform measures, availability of resources and the residents' adaptations to the changed scenario.

All the centrally sponsored schemes like the AMRUT, smart city mission (SCM), Swach Bharat Mission (SBM, Urban), etc., requires increased participation of the Urban local bodies (ULBs) and state governments for their implementation in the form of substantial financial contribution. In addition, the Fourteenth Finance Commission compliance has linked the release of performance grant to the improved physical and financial performance of the concerned state governments and the ULBs [5]. Therefore, the state government's focus is now to strengthen the financial base of the ULBs by its own resource mobilization and to attract and tap the potential external resources through its own capacity utilization and enhancement.

A study [1] on financial inclusions finds that in spite of the increase in banking activities across the talukas in Goa, they are not evenly distributed and show regional variations. They found urbanization, the levels of education and the tourist arrivals as significant factors, and the lower these levels in a region; the lower is its financial inclusion. Thus the smart city concept aimed at strengthening urbanization also holds the key to more financial inclusions.

The SCM of the Government of India have got the approval of a total of Rs. 98,000 crores to develop 100 smart cities and in the following, there will be a rejuvenation of 500 other such cities over the next 5 years. These beacon 100 smart cities were selected through a 'smart city challenge' based on all round social and economic feasibility and impact. In this competition, Bhubaneswar has stood first in January 2016. The mission of a smart city is governed by seven pillars which are the planning and design of the city, its governance (or e-governance) through which it connects with its people, waste management in houses and the city, urban mobility which includes its transport and road connectivity, provisions of shelter to the homeless/slum management, etc., economic development and its socio-cultural development including its health, education, environment, etc.

Migration from rural to urban areas has significant impact on the 'smart city' and 'smart village' objectives [12]. Migration itself is influenced by several 'push factors' like unemployment. Rural poverty, financial exclusions like unavailability of credit facilities, saving opportunities, etc., while the 'pull factors' for migration are better job opportunities, higher wages, urbanized life style, etc.

16.2   CASE STUDY OF BHUBANESWAR

16.2.1   A REJUVENATION

Bhubaneswar the 'temple city' of Odisha with a rich cultural heritage alongside has become the educational and technological hub of the state as well as the country. It was chosen as a smart city from March 2016.

The SCM strategy has been used as a 'challenge cum competition' by the Ministry of Urban Development (MoUD), GOI, which required the cities to effectively use its time and physical resources right from the planning stage to participate in the competition which would enable it to receive huge funding's for its strategy based development. The participation of the people is enabled by a special purpose vehicle (SPV) through enhanced ICT and smart technology use for smart solutions and sustainable development of the city. It is one of the flagship strategies along with AMRUT[1] and Swachh Bharat Mission of the MoUD for providing smart solutions with outcome oriented objectives focused on the need to activize the urban local bodies (ULBs). It is government's way of strengthening the ULBs through power devolution and performance appraisal for transforming the urban landscape. The SCM has become the base for the success of the other urban mission programs of the GOI as mentioned above along with Heritage Cities Development and Augmentation Yojana (HRIDAY) and Housing for all of which Bhubaneswar is a participant.

All these components of SCM are interwoven with each other as depicted in Figure 16.1. To coordinate all the ongoing programs e-governance is an inevitable component which itself is a part of government's digital India program. e-Governance increases the effectiveness of the existing workforce with better connectivity of all stakeholders like government, industries, and the citizens being the major players.

16.2.2   SMART CITY BHUBANESWAR'S MEGA-PROJECTS

This section mentions the several projects of the smart city Bhubaneswar which are either in an 'ongoing' status or 'soon to be implemented' under its smart city initiatives (SCI). Some of these projects are:

1. **The Bhubaneswar Town Center Project (BTC):** This is its initiative of transit oriented development (TOD) under the area based development vision covering 12 acres site which would be the multimodal hub of Bhubaneswar connecting the railway station with its main commercial spine-the Janpath Road.
2. **The Inter-State Bus Terminal Baramunda:** This aims to create a world class terminus covering 20,000 square kilometers area for commercial development increasing the accommodation to approximately 2000 buses from its present capacity of 800 busses on a daily basis.
3. **Slum Redevelopment and Affordable Housing Projects:** This is a PPP model of the Bhubaneswar Development Authority (BDA)

---

[1]Under the AMRUT (Atal Mission for Rejuvenation and Urban Transformation)mission, the government aims to provide some basic services to all its citizens especially the poor like water supply and sewage system, urban transport and other modern amenities to improve the quality of their life.

under the Greenfield project to provide affordable housing for the slum dwellers near the Chandrasekharpur Mouza, covering around 20 acres and some other locations in the city.

4. **The 'Odyssey City Card':** It is a PPP project of the Bhubaneswar Smart City Limited (BSCL) with the ICICI bank to provide card-less payments for an array of services like e-commerce transactions, payment of bills, public transport, etc., easing the lives of some 1 million population of the city.

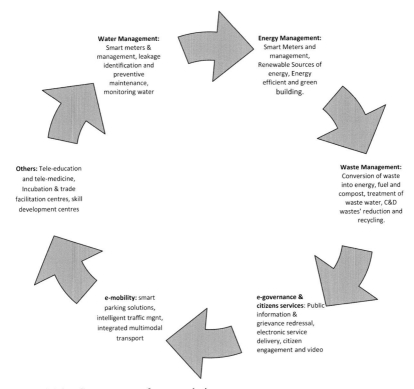

FIGURE 16.1   Components of smart solutions.
Source: Compiled from the data of MoUD, GOI.

16.2.3   *SOME CONCERNS FOR THE SMART CITY INITIATIVES (SCI)*

The SCI has given rise to some concerns as it has suddenly developed a virtual global village with fast fading boundaries. Some of these are:

1. **Employment Aspects:** The automation and digitalization of the production, exchange, and distribution segments of almost all sectors are resulting in replacing the variable human resources with fixed factors leading to loss of job. However, the crux of the situation is to understand that it is the level of skill which is required to be enhanced for full realization of the benefits of digitalization and prevent job dislocation. The problem the government/economy has to handle is 'technological unemployment,' which is not only rendering the unskilled jobless but is depriving the skilled off their job by relatively degrading them vis-a-vis the sudden advent of technology overtaking the existing ones. Thus this requires not only the revamping of the education policies and the curriculum to equip the rising workforce with the needed skill but also requires the up gradation of the existing workforce peeled off by the technology heave.

2. **Migration Aspects:** Migration of people from rural to urban areas has adverse impact on both the regions. The rural areas experiences loss of labor affecting its farm sector development which in turn affects the industrial inputs availability. Apart the cities have their issues of negative externalities from large-scale inward migration like congestion, competition for job opportunities, pollution, slum spread, etc.

3. **Strengthening City and Village Linkages for Mutual Smartness:** The goal of transforming the cities into 'smart cities' cannot be achieved without simultaneously facilitating smart villages due to the several linkages between the two starting from agriculture, financial inclusion, manufacturing, employment, and migration issues, etc., to cite a few instances.

16.2.4   *A BRIEF IMPACT ANALYSIS OF THE ADOPTED SMART SOLUTIONS*

Bhubaneswar is one among the first planned cities of India. It was designed in 1946 by Otto Konigsberger, a German architect. Over time, it has become a hub of educational, health, manufacturing, and service centers like IT, etc., taking pace with the fast developing cities of India and thus could win the Best smart city award.

The city joined the 'smart city club of 100 cities' only in March 2016. So the impact in its wide bloom can only be known after some more years in terms of its overall influence on total income and job generation as the multiplier process would take its lag. This section highlights impacts within this short period of time as is observed from relevant secondary sources:

1. The transit system in the city is updated. Through its G2C (Government to Citizens) initiatives it is providing most services online like motor vehicle tax collections, license applications and approvals, grievance redressal, authorization for national permits for vehicle carrying goods, user friendly transport administration like MO bus apps, E-VCR (vehicle check report), road safety initiative by instant monitoring, tracking, and coordination, GIS mapping of the entire network of road for centralized monitoring of the transport system, etc.

2. Under the Motor Vehicles Act 2019 offenses like drunk driving, not wearing a helmet while driving, over speeding, non-possession of essential vehicle-related documents like pollution check, insurance are punishable offenses with hefty fines. The smart laws have resulted in a rush for regularization like the jump in the license applicants from 50 to 60 per day in the Regional Transport Offices (RTOs) in Bhubaneswar to 2000 per day [15].

3. e-mobility projects: Bhubaneswar's e-mobility projects aims at a 20% increase in public transport by 2021. Its Bus Modernization plan would cover a corridor of 38.7 km requiring some 148 new electric buses to handle a daily passenger load of 192,000. It also proposes deployment of e-rickshaws of about 500 in number along with the required infrastructure including charging stations by 2021. It targets 30% electric operated vehicle travel by 2030 to keep up to its mission of cleanliness maintenance and environmental issues. The overall cost of the project is around Rs.219.75 crores.

4. The digitalization of the tax regime supplementing the goods and services tax (GST) has resulted in an increase in tax collection in the state. It has grown by about 2.3% in Odisha after the tax reforms were adopted.

5. The SCI has opened up vistas of job opportunities in the field of city administration (like of city and infrastructure planner, traffic controller and coordinator, security personnel and experts, budget manager and policy coordinators, etc.), in the domain of software

and hardware technology like (system and data analysts, surveil-lance, and security manager, cybersecurity specialists, hardware specialists and developers, etc.), and in product development (like product manager, product, and service analysts, etc.). Besides the surge of e commerce has expanded job opportunities to far-flung areas both in skilled and unskilled areas.

6. Though not much economic variable figures are readily available as the city has just got into the smart city club, here is presented some developments of Khurdha district in which Bhubaneswar Municipal corporation is located like improvements in education, infrastructure facilities, set up of industries both big manufac-turing as well as medium and small enterprises (MSMEs) to take examples of few(as shown in Table 16.1).

    Some prominent changes within this short period of 2013–14 and 2018–19, i.e., pre, and post period of Bhubaneswar joining the 'Smart City' club is observed of Khurda district. Mentioning some chosen parameters, in higher education an increase of 39% in the number of governments' degree college, in banking sector an increase in investment of 48% in deposits of all scheduled commer-cial banks is seen. Similarly, there has been an increase of 41% in MSMEs with many government schemes to promote skill develop-ment and entrepreneurship amongst the youth in particular and to enable them to become provider of employment rather than seekers. There has been a 52% increase in total investment of MSMEs with a 59% increase in employment over this short period from 2014–15 to 2018–19 in this sector. This is to highlight that the development of the city through its forward and backward linkages leads to overall development in education, infrastructure, income, savings, and investments and to direct and indirect job opportunities.

7. The BMC has started its planned allotment of unique numbers to the houses under its civic jurisdiction to assess and simplify the collection of holding taxes which is one of its major sources of revenue. This will further enable it to geo-tag the numbers and maintain follow-up records which would remain invariant even with the change of tenant status. The BMC has already been able to collect an amount of Rs.40 crores in the fiscal year 2018–19 which of course has fallen short of its targeted collection of Rs.60 crores. The attempt to tap this potential amount this fiscal through a strengthened and streamlined approach.

TABLE 16.1 Some Economic Parameters of Khurdha District 2015–2019

		Education			
Number of Primary Schools	**Percentage Change**	**Enrolment No.**	**Percentage Change**	**No. of Govt. Degree Colleges**	**Percentage Change**
1020 (2013–14)	3.7%	189473	1.6%	43	39%
1058 (2017–18)		186531		60	

		Infrastructure			
No. of Villages Electrified	**Percentage Change**	**All Scheduled Commercial Banks No. of Offices**	**Percentage Change**	**Deposits (in crores)**	**Percentage Change**
1343 (2013–14)	1	570 (2015)	11	529862 (2015–16)	48
1356 (2018–19)		632 (2019)		784705.3 (2018–19)	

		MSMEs			
Number of MSMEs Set Up	**Percentage change**	**Total Investment (in lakh)**	**Percentage Change**	**Total Employment Generated**	**Percentage Change**
2187 (2014–15)	41	10870.89 (2014–15)	52	7176 (2014–15)	59
3085 (2018–19)		16483.84 (2018–19)		11419 (2018–19)	

Source: District at a glance (various years).

All these developments have the potential to address the issues of poverty, unemployment, migration, etc. positively, as it connects the individuals with his social, political, and economic environment through the smart solutions of the smart city. The G2C (Government to citizens) and C2C (citizens to citizens) connectivity made possible by the wave of technology has further strengthened the democracy with every issue getting overall public vision and opinion.

16.3 CONCLUSIONS

It is to be understood that while the revolution of digital technology is a global phenomenon, the pace at which they are adopted and adapted in the very culture of life and work varies at the national and regional level, which largely depends on their economic, political, social, and cultural factors. The smart city solutions involve the dimensions of retrofitting(which refers to identifying the area-specific deficiencies and rectification through effective intervention), redevelopment (covers those areas which are not amenable to any interventions and needs reconstructions), and greenfield development (which refers to the development of earlier uncovered areas with the smart initiatives from the beginning) aimed at sustainable development, increased employment and income, affordable houses, technical connectivity, in a nutshell, an improved quality of life. The 100 cities chosen under this scheme would act as beacons to be followed by the other aspiring cities. In this way, the whole country can stand up smart to compete and triumph in the global challenge of growth and development.

KEYWORDS

- **employment smart city**
- **information technology**
- **smart city mission**
- **special purpose vehicle**
- **transit-oriented development**
- **urban local bodies**

REFERENCES

1. Bawa, M., & Sudarsan, P. K., (2018). Financial inclusion and its determinants: The case of Goa. *Journal of Economic Policy and Research, 13*(2), 2–16.
2. bda.gov.in/project-details/12 (accessed on 3 November 2020).
3. Bhubaneswar e-mobility Plan, Submitted by Bhubaneswar-Puri Transport Services Limited, 2018, http://smartcities.gov.in/upload/smart_solution/5a277bcb24008BH UBANESWAR%20E-MOBILITY%20PLAN.pdf (accessed on 19 February 2020).
4. District at a Glance. Government of Odisha, Directorate of Economics and Statistics, Odisha (various years), http://www.desorissa.nic.in/latest_publications.html (accessed on 19 February 2020).
5. *Draft Financial Management Action Plan (FMAP).* Bhubaneswar Municipal Corporation, prepared for Housing and Urban Development Department, 2016.
6. Jeena, Z., (2017). Insight into the Incipient Smart Cities Phenomena in India. https://smarttech.gatech.edu/handle/1853/58547 (accessed on 3 November 2020).
7. Muhleisen, M., (2018). *The Long and the Short of the Digital Revolution.* Finance and Development.
8. Odisha Sees 2.3% Growth in tax Collections after GST, Press Trust of India, Bhubaneswar, 27th January, 2018, accessed at https://www.thehindubusinessline.com/news/national/odisha-sees-23-growth-in-tax-collections-after-gst/article9981739.ece (accessed on 19 February 2020).
9. Promoting Innovative Smart Solutions: Smart Cities Mission, AMRUT and Swachh Bharat Mission, MoUD, GOI, accessed at http://smartcities.gov.in/upload/smart_solut ion/5954cc108b07cInnovation_for_Smart_Solutions%20FINAL.pdf (accessed on 19 February 2020).
10. Smart Cities Mission (2019). Ministry of Housing and Urban Affairs, GOI, November 14th, 2019, accessed at https://www.commonfloor.com/guide/smart-cities-mission-vs-amrut-a-comparative-analysis-52192.html (accessed on 19 February 2020).
11. Smart City Bhubaneswar: A Citizen Centered Approach for Smart Growth. Accessed at: www.https://hub.beesmart.city/city-portrait/smart-city-portrait-bhubaneswar-india (accessed on 19 February 2020).
12. Srivastava, P. (2015). Rural Urban Migration: Disturbing the Equilibrium Between Smart Cities and Smart Villages, *FIIB Business Review, 4*(3), July–September, 2015.
13. Vaidya, Chetan (2015). Urban Planning At The Core of Smart Cities, accessed at https://egov.eletsonline.com/2015/08/urban-planning-at-core-of-smart-cities/ (accessed on 19 February 2020).
14. www.icicibaank.com (accessed on 3 November 2020).
15. www.newindianexpress.com, Indian Express (accessed on 3 November 2020).
16. https//timesofindia.indiatimes.com/city/bhubaneswar (accessed on 3 November 2020).

CHAPTER 17

An Overview on Blockchain Applications and Consensus Protocols

RAJAT RAJESH NARSAPUR, SHRESHTHA GHOSH, POULAMI BOSE, RISHABH ARORA, MEET K. SHAH, and PRASANT KUMAR PATTNAIK

School of Computer Engineering, Kalinga Institute of Industrial Technology, Deemed to be University, Bhubaneswar, Odisha–751024, India

ABSTRACT

This chapter discusses blockchain technology, keeping in view of its ability to store information with digital signatures in a decentralized and distributed network. Data consistency, transparency, user privacy, these are a few of the basic features which can be utilized in the supply chain, financial, and social services, risk management, healthcare, law, and many more. This chapter explains the taxonomy and architecture of blockchain, and aims to conduct the survey, identify, analyze, and organize the literature on blockchain in supply chain management (SCM) and industrial applications. In addition, the goal is to inspect the proof of stakesible enhancements that blockchain would provide including major disruptions and challenges that arise because of its adoption.

17.1 INTRODUCTION

Blockchain was first invented by Satoshi Nakamoto which was the first decentralized and public transfer ledger in the world. Later, several other use cases of blockchain emerged. Blockchain technology has the potential to become omnipresent in supply chain, financial services, and industries. Recent growth in technologies, mainly associated with Industry 4.0, are provoking the supply chain management (SCM) field to develop new business models with Blockchain technology. SCM with blockchain is a well-researched and proven strategy that has a proof of stakeitive impact on companies' overall performance. It enables companies to achieve different advantages such as reducing cost, improving service level, as well as reacting faster and more efficiently to changes. In general, block-chain allows transferring transactions safely between two or more parties in a digital decentralized ledger without the need of intermediaries [5].

17.2   THE THEORY OF BLOCKCHAIN

Blockchain is a special kind of distributed and decentralized database. The transactions are recorded in a digital ledger and multiple transactions form a block. All these transactions take place in a decentralized means which eliminates the requirement of any mediators to validate or verify the transactions [1]. A block may consist of monetary transactions (Bitcoin) or smart contracts (Ethereum).

A verified node initiates a transaction in a decentralized blockchain network by employing a digital signature using private key cryptography. A block is created which represents the transaction. The transactions can be understood as data structures that represent transfer of digital assets between peers on the blockchain network. The block is then broadcasted to all the nodes in the network. An unconfirmed transaction pool stores all the transactions and is propagated in the network by using a flooding protocol known as Gossip protocol. Then, peers choose and validate those transactions based on some preset criteria. This helps to verify whether the initiator has sufficient amounts for the transaction and so on. The verified block is time-stamped and linked to other blocks in the chain. Thus how the transaction between two nodes is completed via blockchain [5].

Hashing is one of the main concepts in blockchain. Hashes are the unique digital fingerprints of a transaction in a block. Each block represents transaction and carries information such as data, hashes, and the previous hashes. Each time a block is created, it generates a hash based on the previous hash in the former block. The first hash is an exception and is called Genesis. New transactions are validated by miners and then recorded on the global ledger. A difficult mathematical problem based on a cryptographic hash algorithm has to be solved by the miner. The solution found is called the *Proof-of-Work*. This proves that the miner did spend time and resources to solve the given problem on its own. No single miner is solely responsible for ledger within the blockchain, this increases the trustworthiness. All the peers in the network then verify the new block using a consensus mechanism, which is a technique that assists a decentralized network come to an agreement on certain matters. Then the new block would be added to the blockchain and the local copy of each peer's immutable ledger is confirmed.

The blockchain can be created over a private or public ledger. The private ledger is developed over centralized architecture, and the public ledger is made over distributed architecture. Blockchain is a distributed ledger. Distributed means that each node in the blockchain contains its own copy of the blockchain. This makes it difficult to manipulate, as the hacker would have to manipulate every single copy of blockchain in the network.

Blockchain can be categorized into three types:

1. **Public (or Permissionless) Blockchain:** These are designed to be highly decentralized which is open source. It allows anyone to include as a user, miners, developers or community members, e.g., bitcoin, ethereum.
2. **Private (or Permissioned) Blockchain:** These or permissioned Blockchain or enterprise Blockchain has only the participants who have the consent to join the network, e.g., Hyperledger, R3 Corda.
3. **Hybrid Blockchain:** This is a result of an intersection of the privacy characteristic of a private blockchain and the security and transparency of a public blockchain [1].

17.3 CONSENSUS ALGORITHM

Consensus algorithm is the mechanism by which Blockchain nodes reach a consensus on the present state of the distributed ledger. They assure that all the protocol rules are being followed and establish trust between unknown peers in distributed computing.

17.3.1 PROOF OF WORK

This consensus algorithm selects a miner who generates the next block. Bitcoin uses this Proof of Work consensus algorithm. The idea of this algorithm is to solve a difficult mathematical puzzle and give out a solution. This mathematical puzzle needs a lot of computational power and therefore, the node which solves the complex puzzle first gets to mine the next block.

17.3.2   PROOF OF STAKE

In this consensus algorithm, the validator invests in the system coins by locking up a few of their coins as stakes. Then all those validators start validating the blocks. Validator's place bets on validated blocks, which they think can be added to the chain. Depending upon actual blocks which are added in the blockchain, the validators get rewards proportional to their bets and their stake increases correspondingly. To generate a new block, at the end a validator is chosen based on the network economic stake.

17.3.3   PRACTICAL BYZANTINE FAULT TOLERANCE

Distributed networks have a feature called BFT, which they use to reach to an agreement even when few of the nodes in the network fail to respond or when they respond with wrong information. Thus, this reduces the risk of faulty nodes.

17.3.4   DELEGATED PROOF OF STAKE

Delegated Proof of Stake is a consensus algorithm on blockchain, where a voting system counts where stakeholders outsource their work to a third party. The delegates or witnesses are responsible for achieving consensus during the generation and the validation of new blocks. The voting power is proportional to the number of coins the user holds.

17.3.5   TENDERMINT

A blockchain consensus engine and a generic application interface are the two main technical components of tendermint consensus algorithm. Tendermint Core, the consensus engine, makes sure that the equal number of transactions is recorded on every machine in the same order. The application interface allows the interface to undergo processing in any programming language.

17.3.6 COMPARISON OF DIFFERENT CONSENSUS ALGORITHMS

Proof of Stake and Delegated Proof of Stake are used as an alternative to the Proof of Work consensus algorithm. The Proof of Work system requires a lot of external resources for designing and a large amount of computational work in order to secure an immutable, decentralized, and transparent distributed ledger. Proof of Stake and Delegated Proof of Stake require fewer resources and are designed more sustainable and eco-friendly.

There are various differences between consensus algorithms corresponding to their properties. Some of the properties and the behavior of consensus algorithms are explained in the subsections.

17.3.6.1 NODE IDENTITY MANAGEMENT

The joining or leaving the Blockchain Network, i.e., Identity Management of a Node, decides whether a node requires a permission to join the network or is open to join and leave without seeking any authoritative permission. Proof of Stake, Proof of Work and Delegated Proof of Stake Algorithms work best in the case of permissionless blockchain. Whereas permissioned blockchains use tendermint and PBFT protocols [1].

17.3.6.2 POWER AND ENERGY CONSUMPTION

Proof of Work consumes high amount of power and energy in order to mine or commit a block while Proof of Stake and Delegated Proof of Stake are more power-efficient. Whereas, PBFT, and Tendermint Protocols save power to a greater extent [1].

17.3.6.3 ADVERSARY TOLERANCE

The ability of a consensus protocol to yield correct results, as in the absence of any adversary, with adversaries (malicious actors) occupying a certain percentage of the total number of nodes is known as adversary tolerance. The Proof of Work algorithm can sustain less than or equal to 25% malicious nodes whereas Proof of Stake, Delegated Proof of Stake, PBFT, and Tendermint algorithms can sustain up to 51%, 51%, 33.3%, and 33.3% adversaries, respectively [1].

17.3.6.4   *SCALABILITY*

Scalability describes the ability of a Consensus Protocol to make the distributed network reach a consensus with increase in the number of nodes. All protocols except PBFT work well with increase in the number of nodes [1].

17.3.6.5   *PERFORMANCE*

Performance is the ability of a consensus protocol to make the distributed network reach a consensus for committal of a transaction. Proof of Work and Proof of Stake can commit less than 25 transactions per second, therefore displaying low performance as a result of taking relatively high amount of time to help reach a consensus. Delegated Proof of Stake can commit around 400 to 500 transactions per second, displaying moderate performance as a result of taking relatively moderate time to help reach a consensus. Whereas, PBFT, and Tendermint can commit around 1000 and 10000 transactions per second, respectively, displaying high-performance ability as a result of taking much lower time to help reach a consensus [1].

17.4   INDUSTRIAL APPLICATIONS OF BLOCKCHAIN

Blockchain took hold as a promising technology with the introduction and wide use of cryptocurrency. However, it is showing signs as a promising technology enabling many applications in various domains. This technology can be implemented for finding solutions for different domains, such as healthcare, voting, identity management, governance, supply chain, energy resources, and so on. Many experts predict blockchain as a game-changer in the lives of all human beings, similar to how the world has changed after the introduction of Internet to the world. From smartphones and text messages to streaming movies and video conferences with loved ones, as well as for attending meetings or interviews, no one knew the ways the world would change with the invention of the Internet. In this section, we discuss several Blockchain industrial applications that provide many business and industrial benefits. We categorize these applications based on their domains [4].

17.4.1 SUPPLY CHAIN

SCM refers to managing the flow of goods, services, and information in an adequate form so as to achieve high performance and decrease risks. SCM is an attempt by the suppliers to develop and implement supply chains that are as much efficient and economical as attainable. Proper implementation of SCM can result in benefits like increase in sales and revenues, decrease in frauds and overhead costs, and quality improvisation. SCM needs proper interconnectivity between different elements of the supply chain. These might become a problem for small companies as when it grows; the interconnectivity of different elements starts to become complex [12].

Blockchain can be applied in SCM to address the issues. Blockchain applications can be a solution for trust issues in supply chains which can thus sabotage the relationship between the company and the customers. Distributed ledger in blockchain provides transparency and traceability throughout the process and thus reduces the need of checks and balances. Overall cost of moving items is reduced with the help of real-time tracking of products in the supply chain [11].

Development and implementations of Blockchain technologies in SCM is still in its early stage. Blockchain has much potential to improve SCM in the future and eliminate vulnerabilities and inefficiencies from the current system. The crux of a SCM system is to keep it robust, visibility, and better communication and effective and timely results [2].

17.4.2 FINANCIAL INDUSTRY

The advent of blockchain in the world took place as an application to cryptocurrency. It was quite evident, with the huge success of Cyprotcurrency to influence potential investors to invest, that blockchain has innumerable advantages in line for ease in the Financial Industry. In general, trusted third parties are used to conduct financial activities among people and organizations. These third parties provide four functions, namely, confirming the reality of trades, avoiding financial transaction duplication, registering, and validating financial activities, functioning as agents between two parties for ensuring support and trust in legal procedures [3].

Blockchain can serve as a replacement to major of these roles. With blockchain it is easy to prevent a client from performing multiple payments with a total amount that exceeds what they owe. Actually, in the current situation, it is Proof of Stakesible to illegally perform this act with multiple cheque issuances. At the same time, blockchain can act as a secure registry for the conducted financial transactions. This registry cannot be modified by any entity involved after being appended to the chain. It can also be used to validate the conducted transactions through collective checks and verification. These two features enable many financial applications such as the following examples.

17.4.2.1   DIGITAL CURRENCIES (CRYPTOCURRENCY)

There are more than 500 digital currencies available worldwide as of January 2015. These currencies are mainly based on blockchain. Blockchain is used to record and validate digital money ownership and to conduct, register, and verify digital money payments. Bitcoin was the first cryptocurrency initially released in 2009. A recent study in 2017 estimated that there were 2.9 to 5.8 million unique users using cryptocurrencies, most of them are Bitcoin users [3].

17.4.2.2   STOCK TRADING

Stock trading is usually done through a centralized authority such as an exchange market that keeps track of all trades and settlements. However, this process is usually associated with extra costs and settlement delays. To avoid these issues, a platform is being designed based on blockchain to reduce costs and settlement time while at the same time increase transparency and auditability. This platform is integrated with cryptographically secure distributed ledgers to facilities settlement processes. Another successful stock trading facility by using blockchain has been setup by "Chain." Chain developed a live Blockchain integration to link NASDAQ's stock exchange and Citi's banking systems [3].

17.4.2.3   INSURANCE MARKETPLACE

Different insurance marketplace transactions among different clients, policyholders, and insurance companies can be done with the help of blockchain. Blockchain can act as an effective tool for registration, negotiation, and purchase of insurance policies. It can be a boon to claimants, as it shall provide a wide area for a transparent platform for processing claims, furthering the reinsurance activities among insurance companies. Also, bringing the process to a digital platform, a trust relationship can be ensured by inducting Smart Contracts-which may contain the Insurance Policy in the form of a programmed code, which allows for the automation of executing the terms and conditions of a policy. It reduces the effort needed and the costs of execution due to which, the cost of the insurance products can be reduced by the insurance companies and be more competitive to attract more customers. At the same time, it allows insurance companies to launch new automated insurance products for their clients without worrying too much about their administration overhead and costs.

17.4.2.4   FINANCIAL SETTLEMENTS

Blockchain can be used among companies and organizations for recording and processing financial settlements without involving a central authority. Blockchain-based financial settlement systems can be integrated with other blockchain-based applications such as blockchain-based logistics applications or blockchain-based stock trading applications to enable settlement processes [3].

17.4.2.5   PEER-TO-PEER GLOBAL FINANCIAL TRANSACTIONS

Usually, any financial transaction performed between people must go through some form of authoritative entity of a financial institution for verification and guarantees. This entity needs to verify finances and ensure accurate execution of these transactions. However, many of these transactions can now be digitized and verified through blockchain. As a result, the intermediate node is cut out from the process and people can collectively verify and ensure correct execution and recording of these transactions [3].

17.4.3   HEALTHCARE INDUSTRY

Blockchain can serve as an effective tool for tracking the supply of drugs and Patient Data Management. Blockchain can serve as a solution to identify the drug sellers of the developing countries because all the transactions added to the distributed ledger are immutable and digitally time-stamped, which makes a go-through for tracking the supply of a drug, especially which may be harmful or illegal to be sold without authority consent, and ensures that the information is consistent and tamper-proof. Blockchain is capable of establishing a robust and secure transparent framework of storing digital medical records that brings quality services for the patients while reducing the treatment cost [8].

17.4.4   LOGISTICS INDUSTRY

Logistics management applications help to manage the delivery of raw material, products, and services between the producers and the consumer. These can all be part of a single organization through blockchain as it can provide powerful support to enable these applications. One of the complexities in logistics management is the involvement of multiple companies in the activities. It is important for any logistics management application to provide a set of functions to plan, schedule, coordinate, monitor, and validate the performed activities [3]. Such functions can be efficiently and securely supported by blockchain. Audit logistics transactions will help to reduce time delays, management costs, and human errors. By applying smart contracts, it will facilitate agreements between the companies involved and create binding the contracts faster with lower costs.

17.4.5   POWER AND ENERGY INDUSTRY

Blockchain is used for validation and recording the transactions associated with power in microgrids. A localized set of electric power sources which is managed with an objective of enhancing energy production is called a microgrid. One of the main advantages of the microgrid technology offers the cream layer of revenue that can be generated by selling away to the grids, the produced power or the excess power generated by factories during their functioning [3].

Blockchain can be used at larger scales to enable energy trading in smart grids. Smart grids are equipped with bidirectional communication flow where blockchain can be used to support secure and privacy maintained consumption monitoring and energy trading without a need for a central intermediary. Blockchain can be used to enable energy trading in the industrial Internet of things (IIoT). Generally, utilizing blockchain for energy-related applications has the potential to reduce energy costs as well as increase resilience [6].

Other applications of blockchain are namely, voting in which blockchain would solve the issue of alteration of votes by providing a distributed ledger that would ensure votes are counted since the a voter owns is the same as the one counting the total, and Identity Management in which blockchain helps in identification of management that could enable the consumer to access and verify online payments by simply using an app for authentication instead of using a username and password or biometric methods [7].

17.5 CHALLENGES OF BLOCKCHAIN

There are many issues concerning blockchain which have been explained in sub-sections.

17.5.1 SCALABILITY

All the applications that require blockchain technology generate huge amounts of transactions to be processed and linked, which could easily degrade the overall performance. Another issue arises when the blockchain is mined to find, verify, or use earlier transactions. This process involves various steps and the performance is inversely proportional to the size of the blockchain. Thus, the bigger the blockchain the slower the process gets. Scalability has become a pressing issue with the rapid growth in the size and number of entities involved and transactions being performed [10].

17.5.2 SECURITY AND PRIVACY

Blockchain technologies used to improve the reliability of security infrastructure and help to improve the security of distributed networks by using anti-malware environment named BitAV, in which users can distribute the virus patterns on the blockchain which is shown in Noyes that BitAV can improve the scanning speed and enhance the fault reliability (i.e., less susceptible to targeted denial-of-service attacks.

17.5.3 SELFISH MINING

Selfish mining is another challenge faced by blockchain. A block is susceptible to cheating if a small portion of hashing power is used. In selfish mining, the miners keep the mined blocks without broadcasting to the network and create a private branch which gets broadcast only after certain requirements are met. In this case, honest miners waste a lot of time and resources while the private chain is mined by selfish miners [9].

17.5.4 INTEROPERABILITY

It is evident that many industries are currently interested in adopting Blockchain technology but they are falling short of a standard protocol for collaboration and integration of their services to generate mutual benefits. This situation can be associated with the term "lack of interoperability." This is the reason that even after having multiple applications and offering cost-effective and efficient solutions, only cryptocurrency has been able to convince the world as a stable and successful example to a Blockchain-oriented application [9].

17.6 CONCLUSION

A concern raised is on the ability of Blockchain meeting Performance and Scalability standards with the increasing demand from Private and Public Sector entities for its applications. In a Blockchain Distributed Ledger, a trusted relationship between the participants needs to be ensured using various consensus protocols, which then makes way for a transaction to be verified and then appended into the ledger. The process is complex, but effective in comparison to other available systems, at least in the presence of a Blockchain of finite size. Above that, cryptocurrency, which are notable Blockchain platforms are facing numerous regularity issues with major Governments banning their use due to the central authority losing control over the administration of economic policies. Blockchain, with the help of various newly developed consensus protocols, can help increase efficiency in various Industrial and Public sectors, given that its issues are addressed effectively.

KEYWORDS

- **blockchain**
- **cryptocurrency**
- **data consistency**
- **data transparency**
- **industrial Internet of things**
- **supply chain management**

REFERENCES

1. Monrat, A. A., Schelén, O., & Andersson, K., (2019). A survey of blockchain from the perspectives of applications, challenges, and opportunities. In: IEEE Access (Vol. 7, pp. 117134–117151).
2. Acungil, S., (2019). *Blockchain Enhanced Supply Chain*. Thesis. Istanbul Technical University.
3. Al-Jaroodi, J., & Mohamed, N., (2019). *Blockchain in Industries: A Survey* (p. 1.). IEEE Access. 10.1109/ACCESS.2019.2903554.
4. Chen, W., et al., (2018). A survey of blockchain applications in different domains. *Proceedings of the 2018 International Conference on Blockchain Technology and Application-ICBTA 2018.*
5. Petersson, E., & Baur, K., (2018). *Impacts of Blockchain Technology on Supply Chain Collaboration*. Master Thesis. Jönköping University.
6. Ali, M. S., Vecchio, M., Pincheira, M., Dolui, K., Antonelli, F., & Rehmani, M. H., (2019). Applications of blockchains in the Internet of things: A comprehensive survey. In: *IEEE Communications Surveys and Tutorials* (Vol. 21, No. 2, pp. 1676–1717). Second-quarter.
7. Mahajan, D., (2019). A survey paper on blockchain technology. *International Journal for Research in Applied Science and Engineering Technology, 7.* 3564–3569. 10.22214/ijraset.2019.5584.

8. McGhin, T., & Kim-Kwang, R. C., & Liu, C., & He, D., (2019). blockchain in health-care applications: Research challenges and opportunities. *Journal of Network and Computer Applications*, 135. 10.1016/j.jnca.2019.02.027.
9. Sandeep, K., Abhay, K., & Vanita, V., (2019). A survey paper on blockchain technology, challenges, and opportunities. *International Journal of Engineering Trends and Technology, 67*(4), 16–20.
10. Scherer, M., (2017). *Performance and Scalability of Blockchain Networks and Smart Contracts (Dissertation).*
11. Supriya, T. A., & Vrushali, K., (2017). Blockchain and its applications: A detailed survey. *International Journal of Computer Applications, 180*(3), 29–35.
12. Svennevik, N. J., & Hua, A., (2017). *Project Thesis Blockchain Enabled Trust and Transparency in Supply Chains.* 10.13140/RG.2.2.22304.58886.

Index

World
  ahead programme, 82
  world wide web (WWW), 18
    programs, 16

X

Xenotime, 92, 93

Y

Yen of Japanese (JPY), 175
Youngsters, 41
Youth internet study survey, 43
YouTube, 35